D042266

The quest for quarks

The quest for quarks

BRIAN McCUSKER

Professor of Physics, University of Sydney

Library - St. Joseph's College
222 Clinton Avenue
Brooklyn, N. Y. 11205

CAMBRIDGE UNIVERSITY PRESS

Cambridge

London New York New Rochelle

Melbourne Sydney

Published by the Press Syndicate of the University of Cambridge
The Pitt Building, Trumpington Street, Cambridge CB2 1RP
32 East 57th Street, New York, NY 10022, USA
296 Beaconsfield Parade, Middle Park, Melbourne 3206, Australia

© Cambridge University Press 1983

First published 1983

Printed in Great Britain by the University Press, Cambridge

Library of Congress catalogue card number: 83-7459

British Library cataloguing in publication data
McCusker, C. B. A.
The quest for quarks.
1. Quarks
I. Title
539.7'21   QC793.5.Q252

ISBN 0 521 24850 7

UP

Contents

111719

Acknowledgements

I acknowledge the very great help I have had from my teachers, first of all my mother and my father, then many other people. In physics I am particularly indebted to Lajos Janossy, George Rochester and Erwin Schrödinger. I have been greatly helped by my colleagues in Sydney, particularly Professors Stuart Butler, Harry Messel, Lawrence Peak and Murray Winn. I appreciate the comments and criticisms of the following who read all or part of the manuscript: Professor Luis Alvarez, Harish Briel, Carolyn Carpenter, Dorothy Carpenter, Samantha McKay and Professor Lawrence Peak.

The manuscript was typed by Ms Barbara Dunn and Mrs E. Shooks, and the drawings prepared by Ms Audrey Castleman and Mrs M. Tallon.

I thank Professor George Rochester for his two beautiful photographs of V-particles, Professor Peter Fowler for the photograph of the first τ particle decay, and Professor Alan Thorndike and other members of the Brookhaven Ω^- group for the first bubble chamber photograph of the Ω^- and the accompanying diagram.

I am indebted to Messrs A. P. Watt Ltd and Macmillan Ltd for permission to quote Sir John Squire's epigram on Einstein.

1

The structure of matter

I was born and brought up in north-eastern England. At a fairly
early stage in my scientific education I became acquainted with
the notion of 'atoms' – fundamental, indivisible particles out of
which all things are made. Oddly enough, the first occasion that
I can remember when the idea cropped up was my reading a
newspaper story about a forthcoming attempt to 'Split the
Atom' – to divide the indivisible.

The story was a bit scary. The reporter wondered if a
successful attempt would let loose such vast energies as might
wreak havoc with the world. Scientists were said to pooh-pooh
the idea, and in the short run they were right. But maybe the
intuitive reporter knew better in the long run. The Cockcroft–
Walton experiment was part of the Cambridge research
programme leading to the discovery of the neutron, that in its
turn led to the fission of uranium and that to the atomic and
hydrogen bombs.

Anyhow, the idea of atoms seemed to me very sensible – all
these many complicated things in the Universe neatly explained
as combinations of a few simple objects. It was an old (Grecian)
idea made modern and scientific by John Dalton, a fellow north
country man. When I began to study chemistry at school,
atomism was taken as obvious by my teachers and even more
so by me:

$$2H_2 + O_2 \rightarrow 2H_2O.$$

It was simple, obvious, beautiful.

I heard that only a minority of the ancient Greek philosophers
were atomists and I concluded that most Greek philosophers
must have been rather stupid. This was reinforced when I heard
that they had such an aversion to experiment that they would

argue for days about the number of teeth a horse had rather than examine a horse. I loved experiment; chemistry and physics practical classes were the best times of my school week. There is a story that Wolfgang Pauli, in the early days of quantum mechanics, once burst in upon Heisenberg and Bohr and said 'You should read more of Einstein; he's not half as stupid as you might think!' And, of course, the Greek philosophers were not half as stupid as I thought. Thales, Anixamander, Heraclitus, Parmenides, Pythagoras, Plato, Aristotle; none of them were atomists. Democritus was very much in the minority. So, when John Dalton suggested that matter is atomic, that the material Universe is composed of just a few types of indivisible, fundamental entities, he was going against the weight of informed opinion, established over many centuries.

Dalton had a few very great men of more recent times on his side. Newton favoured atomism.

> It seems probable to me, that God, in the Beginning form'd Matter in solid, massy, hard, impenetrable, moveable Particles, of such Sizes and Figures, and with such other Properties, and in such Proportion to Space, as most conduced to the End for which he form'd them.

Leibniz also tried to erect a doctrine of the world based on 'physical' atoms. But though Dalton had such people of the highest quality on his side, the great weight of opinion was against him.

What drove him to revive this Grecian minority opinion? Mostly the recent (recent in AD 1805) discoveries in chemistry. These could be stated as five laws:

(1) The law of conservation of mass: in a chemical reaction the total mass of the various substances at the end of the reaction is equal to the total mass of the substances at the beginning of the reaction. This had been a difficult law to arrive at because quite often one of the components in a reaction (either before or after) is a gas and people found it difficult to weigh gases – or even, sometimes, to know that they were there. For instance, when iron rusted they did not at first know that it was taking oxygen from the atmosphere.

(2) The law of constant proportions: all pure samples of compounds contain the same elements combined in the same proportions by weight. By this time people had some idea of elements and compounds. Water, for instance, had been separated by Sir Humphry Davy (and an electric current) into hydrogen and oxygen. The hydrogen and oxygen had then been combined (by burning one in an atmosphere of the other) to form water.

(3) The law of multiple proportions: if two elements combine to form more than one compound then the several weights of A combining with a fixed weight of B are in simple ratios. Good examples of this are the oxides of nitrogen, which we would today write as N_2O, NO, N_2O_3, NO_2 and N_2O_5. So if we had 30 units by weight of oxygen then the amounts of nitrogen by weight in these five compounds would be in the ratios $60:30:20:15:12$.

(4) The law of reciprocal proportions (sometimes called the law of equivalents): the weights of elements A, B and C that combine with a fixed weight of another element are the weights with which A, B and C will combine with each other, or simple multiples of them.

(5) Gay-Lussac's law (put forward in 1808): when gases react they do so in volumes which bear a simple ratio to each other and to the volumes of any gaseous products.

It is fairly obvious that most of these laws are nicely explained by the hypothesis that all matter is made from a fairly small number of elements and that the elements are made from atoms, the atoms of any one element being indistinguishable from one another but different to the atoms of any other element. The atoms are considered to be 'true' atoms, indivisible and indestructible.

The first law follows at once; there are just as many of the unchangeable, indestructible atoms at the end of the reaction as there were at the beginning. Laws 2, 3 and 4 are readily seen once we use the modern symbolism: H_2O for water, for instance, meaning that two atoms of hydrogen and one of oxygen give a molecule of water; H_2O_2 for hydrogen peroxide, and so on.

Gay-Lussac's law gave Dalton some trouble. He got round it, in effect, by saying that Gay-Lussac's experiments were wrong – which caused some amusement amongst the cognoscenti, for Gay-Lussac was one of the best experimentalists of the day and Dalton, definitely not so. But it was all tidied up by Avogadro when he drew the distinction between atoms and molecules and realised that elements can come in molecular form. This he did in 1811 though, very surprisingly, his work was not widely recognized till 1860 when his compatriot Cannizzaro drove the point home at a conference in Karlsruhe. By this time the conference delegates could read Cannizzaro's paper as they returned home to their various towns in Europe by steam train rather than the stage coaches of Dalton's day.

By this time, Dalton's idea was being applied in physics, particularly in what is called the kinetic theory of gases. This is the theory that explains (and predicts) the properties of gases by assuming they are made of molecules, which are in rapid, random motion, and that the kinetic energy of motion of the molecules constitutes the heat energy of the gas. This idea seems to have originated with Herapath, who supposed the atoms to be perfectly hard. It is difficult to envisage the collision of perfectly hard spheres – one feels that something has to give, but what? Waterston, who followed Herapath, chose to consider perfectly *elastic* spheres for his molecules. And of course this brought the criticism that he was explaining the elasticity of gases by assuming they were made of elastic particles; and where was the advance? However, Maxwell and Clausius put aside this objection (for the time being) and· developed a mathematical theory of gases, which accounted for many properties of gases and correctly predicted others. From this time on the atomic theory was established in the scientific community. This is not to say that all scientists believed in the reality of atoms, or that they all became 'atomists' rather than 'continuists'. But the majority realised that here was a reasonable model, with considerable predictive power and scope for development. (And quite a few believed wholeheartedly in the atomic hypothesis.)

However, the need to have a perfectly elastic atom to make

the theory work *did* imply that the atom is complex, not the simple, indivisible, particle that is the fundamental building block of all matter. Another line of reasoning suggesting the same thing was the *number* of different sorts of atom needed to explain chemical activity. The ancient Greek atomists had only postulated four (or sometimes five) different sorts: earth, air, fire and water (plus, possibly, aether). But once Mendeleev had systematised his table of the chemical elements, it was clear that at least 92 different types of atom were needed in the 'modern' system. But 92 seemed a large number of allegedly 'fundamental building blocks' of nature. The atoms fell under suspicion of being complex objects. In the last decade of the 19th century this was shown indeed to be so by experiments in a fairly new field of study in physics, the conduction of electricity through gases.

The study of this phenomenon depended on two advances in technique. These were the ability to produce and maintain a 'vacuum', that is, a considerable reduction in gas pressure in a glass vessel of reasonable capacity – a litre, say – and the ability to produce moderately high electrical voltages (a few thousand volts) from a supply that would give a noticeable (if rather small) current. A sketch of the sort of apparatus used is shown in fig. 1.1. The evacuated vessel has two electrodes; that attached to the negative side of the supply is the cathode, the other is the anode. If a small hole is made in the anode one finds that some entity ('cathode rays', they were called) will travel through the hole and, if the far end of the tube is coated with a phosphor, will cause a small bright glowing spot to appear. A modern television tube is a sophisticated descendant of this simple device. It was found that these cathode rays could be deflected in a magnetic field; they acted like a stream of negative electricity (which is exactly what they are).

In Cambridge, J. J. Thomson (later Sir J. J. Thomson, Nobel Laureate) was able to measure the ratio e/m of their charge e to their mass m. A little later it became possible to measure the same ratio e/m for the 'positive rays' moving in the opposite direction in a tube containing a small pressure of

hydrogen gas. This ratio was about 2000 times smaller than the
first ratio e/m but very close to the value obtained by a study
of the electrolysis of hydrogen compounds like hydrochloric
acid. Thomson drew the correct conclusion, namely that he was
dealing with two constituents of the hydrogen atom, a positive
particle, which accounted for most of the hydrogen mass, and
a negative particle of equal and opposite charge and about 2000
times lighter than the positive particle. The negative particle was
called an electron (first by Johnston Stoney, an Irish physicist
who suspected its existence from studies of electrolysis). The
positive particle became known as a proton.

With this simple apparatus, which nowadays any handyman
with a reasonable skill in glassblowing could make, the 'atom'
had been split. It does not take much energy to do it; 100
electron volts is ample for one atom – that is, the energy an
electron acquires when it experiences a potential drop of 100
volts. But it is a much bigger energy per atom than one comes
across in the usual sort of chemical interaction.

So, from the beginning to the end of the 19th century, atoms

Fig. 1.1. A sketch of a gas discharge tube. A vacuum pump is
used to remove most of the gas in the tube, which can then be
sealed off with the tap. An electrical potential of a few thousand
volts is then applied across the two metal electrodes – the
right-hand electrode in the figure is positive. Cathode rays
(electrons) then stream from left to right. If a small hole is made
in the anode some of the electrons can stream through it and
strike a screen of some scintillating material on the glass end of
the tube. A TV tube is a sophisticated descendant of this device.

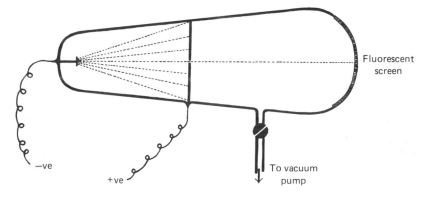

had been suggested, had been made scientifically respectable, and had been shown not to be atoms! Fortunately for the atomists, two new 'particles' had appeared, the electron and proton, which were certainly constituents of the chemical 'atoms' and might be the real fundamental building blocks of nature – the real atoms. And there were only two of them; the modern atomists were (at least temporarily) two up on their colleagues of ancient Greece.

Of course the study of the structure of matter did not stop here (otherwise there would have been no quest for quarks). The idea that matter is made up of just two sorts of particle, one a charge of positive electricity and the other a charge of negative electricity, is simple and appealing but it led to difficulties. To see what these were we need to look at two areas of physics.

The first was fairly well established in 1900. I have mentioned it already: the kinetic theory of gases. It took as its starting point the belief that all gases are made of molecules, and sought to relate the microscopic properties of these molecules, their speeds, diameters, masses and so on, to the macroscopic properties of the gas. It was (and is) very successful. The other branch of physics was quite new in 1900. It was radioactivity. Kinetic theory (as I shall show in a moment) gives us good estimates of the *sizes* of atoms and molecules. Radioactivity shows that, if atoms are the size claimed by the kinetic theory, then most of their volume is 'empty space' and the greater part of their mass is concentrated in a very much smaller volume – just as the greater part of the mass of the Solar System is concentrated in the Sun. From this stems a whole variety of ideas about matter.

The idea redeeming Dalton's theory was Avogadro's hypothesis and the distinction he made between an atom and a molecule. Avogadro realised that two similar atoms, for example two atoms of hydrogen, might combine to form a molecule of hydrogen and that this is the stable form of hydrogen under the conditions holding at the Earth's surface. (In interstellar space things are different. There the usual form of hydrogen is atomic hydrogen.) Avogadro's hypothesis was that equal volumes of gases (at the same temperature and pressure) contain the same number of molecules. The hypothesis is now so well confirmed by experiment that it is often called Avogadro's law. From this law we see that the masses of two equal volumes of different gases are in the ratio of their molecular masses. Chemists quickly realised the usefulness of this in calculations. For any substance, the amount of it that weighs in grams the same as its molecular weight always contains the same number of molecules. They called this amount a gramme-mole. They (and the physicists) found a variety of ways of counting the number of molecules in this amount of matter. This number is now called Avogadro's number.

One of the ways, which is easy to explain, relied on electrolysis – the break up of a compound when an electric current is passed through it. When some substances, such as hydrochloric acid (HCl), are dissolved in water they dissociate. The molecules of HCl dissociate into positive hydrogen ions (an ion is an electrically-charged atom or molecule) and negative chlorine ions. If we dip two electrodes into such a solution and maintain one positive and the other negative, then the positive ions are attracted to the negative electrode (or cathode) and neutralised. They then grab the nearest other neutral hydrogen atom and form a hydrogen molecule. Lots of molecules unite to form a bubble of hydrogen gas, which rises to the surface and can be collected. We need, of course, to supply electricity constantly to keep the cathode negatively charged. We can meter the amount of electricity supplied and so find out how much electricity (that is, what amount of electric charge) is needed to liberate a grm-mole. The smallest unit of electric charge has been measured in various ways – the first successful experiment was carried out by the American physicist R. A. Millikan – and so the number of molecules in a grm-mole became known. It is big, 6.023×10^{23}.

Now a grm-mole of hydrogen gas can occupy a large volume. This volume depends on the temperature and pressure. The gas is usually easy to compress, so one guesses there is a good deal of space between the molecules. But when the hydrogen is liquefied it is difficult to compress any further, so the molecules must be almost touching. By dividing the volume of a grm-mole of liquefied hydrogen by 6.023×10^{23} one gets the volume of one molecule of hydrogen (to a close approximation). From this one gets an idea of its linear dimensions. Most atoms turn out to have radii of about 10^{-10} metres. Once the volume of a molecule is known its cross-sectional area becomes calculable. Then, if the number of molecules in a given volume is known, it is easy to calculate the mean distance a molecule will travel between collisions, which is called its mean free path. Again, by measuring the amount of heat it takes to bring a grm-mole of gas from close to the absolute zero of temperature ($-273\,°\text{C}$) up to, say, $15\,°\text{C}$, one measures the amount of energy in the grm-mole and hence gets the mean kinetic energy of a molecule at $15\,°\text{C}$ (or any other temperature one chooses). The kinetic energy is given by $\frac{1}{2}mV^2$ where m is the mass of the molecule and V its velocity, and so we get an idea of the speed at which the gas molecules are flying around in the atmosphere. At $15\,°\text{C}$ it is about 0.5 km/sec, 1800 km/hr – well over the speed limit! So, by a few quite simple experiments we have found out the number of molecules in a given amount of matter, their size, their mass, and their mean velocities at room temperature.

Now for the other half of the study – radioactivity. In a way, this came out of the study of the conduction of electricity through gases that I spoke of earlier. Röntgen, in 1895, had found that when cathode rays hit a solid object, the glass of the tube for instance, they give off an invisible radiation, which he called X-rays. These X-rays, as you know, can penetrate solid matter and blacken photographic emulsion even when it is in a light-proof packet. Following up this discovery when it was the talk of the world, Becquerel in France discovered that a sealed box of photographic plates was blackened when placed next to a uranium salt. He drew the correct conclusion: the uranium was giving off rays, like X-rays. Radioactivity had been discovered. Both Röntgen and Becquerel were awarded Nobel Prizes for their discoveries. The amazing phenomenon of radioactivity was soon the subject of many investigations. It was spontaneous: a piece of uranium (or thorium) did not need any electric or magnetic field to produce these radiations. They just poured out of it.

Radioactivity stirred the imagination of many people other than professional scientists. In 1905 H. G. Wells wrote a science fiction novel in which he forecast the use of 'atomic bombs'. It is very interesting to compare that prediction, forty years before the atomic attack on Hiroshima, with Lord Rutherford's remark in 1936 that 'anyone who talks of energy from the atom is talking moonshine'. The intuition of the artist is often much more effective in prediction than the detailed knowledge of the 'scientist'.

But Rutherford was a superb experimental investigator in the early days of radioactivity and attracted some very talented physicists from many countries to his laboratories, first in Canada, then Manchester, and then Cambridge. He was one of the people who sorted out the three types of radiation coming from radioactive materials. These were first, α-particles, which turned out to be the nuclei of helium atoms, four times as heavy as protons and with twice their positive electric charge; secondly, β-particles, which turned out to be high speed electrons; and thirdly, γ-radiation, which was like X-radiation but more penetrating than any X-radiation one could generate in those days.

Rutherford found that he could use the α-particles from radium as a kind of probe. He fired these α-particles (whose energy was around 5 million electron volts, written, 5 meV) at thin gold foils. Many of the particles went straight through without noticeable deflection, a few were deflected slightly, and very occasionally one was deflected through a large angle. Now the cause of the deflection seemed almost certain to be the electric charges in the atom. The law of force between two electric charges, q_1 and q_2, is simple:

$$\text{force} = Kq_1q_2/r^2,$$

where K is a constant, which depends on what units of measurement we use and r is the distance between the charges. But if the charge on the atom is spread out over a radius of

10^{-10} m (small as that is) this force is nothing like strong enough to deflect the fast, heavy, α-particles noticeably. Rutherford said that to see these α-particles bouncing back from the gold foil (as they very occasionally did) was as surprising to him as firing a 15-inch shell at a sheet of paper and having it bounce back. For the α-particles to behave in this way the positive charge in the gold atoms must be concentrated in a region about 100 000 times smaller in radius than the whole atom, that is, within a radius of about 10^{-15} m. The idea of the nuclear atom had appeared!

Rutherford pictured the atom as something like a miniature solar system with the tiny, heavy, positively-charged nucleus at the centre, as the Sun is at the centre, and the electrons revolving round it. But this model had its problems. First, the matter in the nucleus must be exceedingly dense, much more dense than lead, in fact about 4×10^{13} times denser! Then there was the odd fact that the nucleus of helium is four times heavier than a proton and had only two positive charges. It looked as if the helium nucleus is made of four protons *and* two electrons. But the helium atom still had its two electrons 'in orbit'. Why should some electrons go into the nucleus and others go into orbit? And the orbital electrons had their difficulties. An electron moving in a circle is being accelerated – its direction of motion is constantly being changed from the straight line that it would follow if no external forces were acting on it. It is just the same for the Earth, of course. Our direction of motion is constantly being changed by the gravitational pull of the Sun. But when an electric charge is accelerated, it radiates energy as electromagnetic radiation, light, radio waves, or X-rays, depending on the energy. In any case it radiates energy, which is to say, it loses energy; it slows down and this makes it spiral in towards the central positive charge. Moreover, with such small distances as we have in the atom, this should happen very quickly – in about 10^{-12} s, a million millionth of a second. So, according to the physics of the day (about 1912), such an atom was not even approximately stable.

Fortunately, at this time, there was a young Danish physicist of genius working in Rutherford's laboratories in Manchester – Nils Bohr. He was aware of the fairly new quantum theory and promptly applied it to this 'planetary' model of the atom. He 'quantised' the orbits of the electron, supposing only certain orbits are allowed. This made the atomic 'planetary' system very different to our Solar System. Moreover, he supposed that when an electron was in one of these orbits it did not lose energy (for some unknown reason). It only lost energy when it jumped (again, in some unknown way) from one orbit to another. This model would not have been accepted if all it did was to explain Rutherford's scattering experiment. But it did much more. With it Bohr was able for the first time to calculate the wavelengths of the light emitted by hydrogen atoms and so to explain the spectrum of hydrogen. He could also make quite a reasonable shot at the spectra of some other elements. This was a spectacular advance.

You will notice that Bohr had to put up with a lot of uncertainty to get his model. He could not say (at this point) why the electrons in stationary orbits did not radiate, or why the nuclear matter was so dense, or what happened when the electron made a jump and so on. He had not by any means got a complete theory of the atom. But he had got a theory that was much better than anything else around. You might also notice another thing. Trying to find the truth about the detailed structure of matter is like trying to catch a well-oiled eel in a darkened aquarium with, maybe, the thought in your mind that there is possibly a piranha or two in the tank.

To come back to the Bohr model of the atom: the study of the fine structure of matter, the search for the fundamental building blocks of the Universe, divides at this point. The two paths seem to go off in different directions, but discoveries on each path have been essential to the study of the other. And later on the paths in a sense rejoin one another. One path is the study of the electrons; the other, of the nucleus.

The first and most obvious difference between them is one of scale. In the Bohr model the nucleus is *very* small, 10^{-15} m in radius, is very dense, having almost all the mass of the atom in it, and is positively charged. The electrons are spread out over a much larger volume (100000 times the radius of the nuclear radius, so 10^{15} times the volume), are quite light particles, and are negatively charged. It was the study of these particles and the outer sections of the atom they occupy that first led to great advances. Almost literally, the nucleus was a tougher nut to crack.

To see what went on in the outer atom we need to backtrack a little to the end of the 19th and beginning of the 20th century. In an address to the British Institution at the turn of the century, Lord Kelvin, a great Scots theoretical physicist, said that physics in the 19th century had made spectacular progress. It was in extremely good shape. Only two small clouds marred the sunny blue sky and he expected these would soon disappear. One was the odd result of Michelson and Morley's experiments on the velocity of light: it appeared to be unaffected by the motion of the Earth around the Sun, contrary to the forecasts of physics. The other was the failure of theoretical physics to account for the way energy radiating from a hot black body varies with wavelength. Now I am sure a lot of people would find the first of these a rather esoteric detail to worry about, and the second one almost ludicrously so. It says a lot for Lord Kelvin's intuition that unerringly he had spotted two major defects in the physics of his day. But his reasoning was not as good as his intuition; these two clouds did not fade away. The first led Einstein to his theories of relativity, to $E = mc^2$ and so on. The second, which is the one that concerns us now, led Max Planck to his quantum theory.

Until Max Planck, physicists had assumed that when energy was transferred from one body to another it could be transferred *continuously*, in arbitrarily small amounts. Planck decided to try the opposite idea, that it could only be transferred in 'packets' of a definite size. He decided that if he were considering electromagnetic radiation of a definite frequency v (nu, the Greek n), then the size of the energy packet was hv, where h is a fundamental constant that is now called Planck's constant. In a way this was applying the atomic hypothesis to energy transfer. But there was a difference. As long as we stick to frequency v, then the size of the packet is hv. But we can always take another frequency, change v and so change hv. We can always, for example, move from bluish-green light to greenish-blue light. But once we fix v, we just cannot make energy transfers of $\frac{1}{3}hv$, or $\frac{11}{5}hv$ – it has to be done in units of hv. It is a bit like currency. If you are paying in bank notes you can only pay in, say, units of \$1. The value of the dollar against gold may be sliding up and down from day to day, but you can only pay in these units.

With this apparently simple modification Planck was able to derive a theoretical 'energy against wavelength' curve that accurately matched the experimental result. And Planck knew that the apparently simple modification was of a very fundamental nature. In fact it led to the overthrow of Newtonian physics and to changes in our ideas about the Universe that are certainly still going on and may well have only just begun.

One of the first people to appreciate the importance of Planck's step was Einstein. In 1905 he used it to explain the photo-electric effect. It had been found that when some metals are illuminated they emit electrons. Some details of this process were quite puzzling until Einstein used the idea that the light energy is delivered to the metal in these 'packets'. The idea explained the formerly puzzling aspects; Einstein was awarded the Nobel Prize principally for this work rather than for his theory of relativity. It was, indeed, a big step. It showed physicists that light, which for nearly a century they had known to be wave-like, behaved, when it interacted with matter, more like a stream of tiny birdshot. Light, in fact, had a dual nature. It diffracted, refracted and interfered like waves do; but when it interacted with, say, a photographic film, it behaved like a stream of particles. These particles were soon called photons.

At first sight this duality seemed impossible. For physicists, particles and waves were a good deal more different than chalk and cheese. Both ideas, of course, were abstractions from reality. But wave motion was thought of as a non-local phenomenon. A system of plane waves is easiest to deal with if thought of as stretching to infinity both forward and back and from side to side – and also as having no beginning or end. A particle, on the other hand, is precisely located. The particle is thought of as a speck of matter; the wave is a wave in some medium. For light, the medium was at first called the aether; later, light was thought of as a wave motion in the electromagnetic 'field'. In practice, a billiard ball is a good approximation to a particle – as is a planet, considered as part of the Solar System. The easiest waves to visualise are probably waves on the surface of a pond or the ocean. The two, particle and wave, are the antithesis of each other, but this new theory was saying that light is *both* – or at least behaves like both from time to time.

Worse was to come, in the sense of more paradoxes. In Paris, in 1925, Prince Louis de Broglie thought that if a wave (like light) might display particulate behaviour, maybe a particle, like an electron, might have also a wave nature. He considered the Bohr atom as a resonating system. The simplest analogy is a guitar string. A guitar string can vibrate in various ways. Some of them are shown in fig. 1.2. Always one must have nodes at the two ends; there, the string is fixed and cannot move. If these are the only nodes then the string is vibrating in what musicians call its fundamental mode. The wavelength of the vibration is

Fig. 1.2. Three possible modes of vibration of a stretched string.

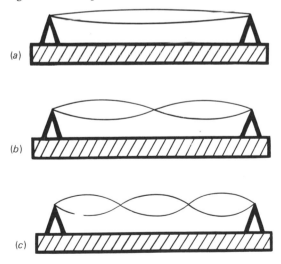

then just twice the length of the string. If we have an extra node in the middle we have the first harmonic; its wavelength is just the length of the string. Then there are higher harmonics, with shorter wavelengths. In a guitar, the string is 'coupled' to the guitar body, which can reinforce some of the harmonics and suppress others. Also, a good guitarist knows how to excite different harmonics by his playing and so produce different tonal qualities at will. De Broglie thought of the Bohr atom as such a system, with various possible harmonics corresponding to the radii of the supposed orbits of the electron. He found he could get the right values if he supposed that the wavelength, λ, to be associated with the electron is given by $\lambda = h/mv$, where m is the mass and v the velocity of the electron, so mv is the *momentum* of the electron. h is Planck's constant, cropping up here once again.

This suggestion that electrons have waves associated with them was radical. But it was also quite easy to check up on. Waves produce characteristic effects – diffraction, interference, and so on. If we have a point source of light and a screen, and we interpose between them an opaque sheet with one pin hole in it, then the image we get on the screen is not just a single bright dot. Instead, it has a number of bright and dark rings around it. The experiment is just a little tricky to do because the wavelength of light is so short, but is standard in physics teaching laboratories. This pattern is due to the diffraction of the light waves. Now, by 1925, it was quite easy to get a stream of electrons. J. J. Thomson had done it in the 1890s when he discovered the electron. In 1926 his son, G. P. Thomson, tried the effect of shooting a beam of electrons through a thin gold foil – and got a diffraction pattern! Like his father, he was awarded a Nobel Prize (in 1937, shared with Dr C. J. Davisson from the USA who had carried out a similar experiment at the same time, with the same result).

So, not only had light a particle aspect, but electrons had a wave aspect. This idea of de Broglie catalysed the development of quantum mechanics. In 1925, Erwin Schrödinger used de Broglie's idea to write his famous equation and produced an accurate description of the hydrogen atom. Werner Heisenberg, using a different mathematical method, got a similar result. The two mathematical methods were shown to be identical by Schrödinger. They became known jointly as quantum mechanics. This new mechanics worked beautifully in the atomic domain, where the old Newtonian mechanics had been a complete failure. Also the new mechanics shaded into Newtonian mechanics when one moved to larger scale phenomena. But there was a price to pay: even the originators of the new description of nature found it hard to swallow. Heisenberg wrote later

> I remember discussions with Bohr which went on through many hours till very late at night and ended almost in despair; and when, at the end of the discussion, I went alone for a walk in the neighbouring park, I repeated to myself again and again the question: can nature possibly be as absurd as it seemed to us in these atomic experiments?

It is worth having a look at some of the things that Heisenberg found so odd as to be absurd. First was a basic uncertainty in nature. In classical mechanics one sought to predict the momentum and position of a given set of bodies at a given time from a knowledge of the basic laws and the initial conditions (which were the positions and momenta at the starting time). For instance, given initial positions and momenta of the Sun, Earth and Moon, and the law of gravity one might wish to predict the time in the future of various types of eclipse. Heisenberg found that introducing the quantum hypothesis meant that the position and momentum of any particle can never both be known with complete accuracy (at any one time). If Δx is the uncertainty in its position, and Δp its uncertainty in momentum then the product of these, $\Delta x \Delta p$, must always be greater than or equal to a constant. And you will probably have no difficulty in guessing which constant: h, Plank's constant. This is Heisenberg's famous Uncertainty Principle. This principle also applies to other pairs of variables, for instance, energy E and time t. So another way of stating the Uncertainty Principle is $\Delta E \Delta t \geqslant h$.

One way of looking at this principle is to note that whenever an observation is made the observed system is necessarily disturbed; we have to see, or hear, or in some way become aware of the system. So the system is altered by the observation. The experimenter is an essential part of the experiment. 19th-century science had tried to forget this and, as a result, could not handle atomic phenomena (and many other phenomena).

Another way of looking at the Uncertainty Principle is to use the $\Delta E \Delta t \geqslant h$ form of it. If we consider a particular point in space – in a vacuum, to make things simple – and if we then consider a very short time interval, say 10^{-24} s, then although the energy at the point in the vacuum is, over a long time, zero, the Uncertainty Principle says that $\Delta E \Delta t \geqslant h$, and if Δt is very small, ΔE must be big. It might even be bigger than the energy needed to make a pair of electrons (one positive, one negative – we shall come to positive electrons in a short while). So these two will just pop into existence in the void. They can only last for this very short time and then disappear again. But that is happening all over the void, all the time. If Δt is even shorter, ΔE may be big enough to produce a pair of positive and negative protons, and so on. The quantum-mechanical vacuum is not a dead, empty space but a very lively can of worms. Having all those particles appearing and disappearing (in positive–negative pairs) naturally affects the electrical properties of the vacuum. These effects can be predicted and experiment shows the prediction to be correct.

So, uncertainty is built into the Universe. Something like this had been noted earlier by the experimentalists. They found that it was impossible to say when a given radioactive atom would decay. Nowadays one can observe the passage of a single radioactive particle through, say, a bubble chamber. Many pions (we will be coming across these soon) have been seen doing just this. The pion has a moderately short half-life, about one hundred millionth of a second (10^{-8} s). Some pions are seen to decay after quite a short flight, but some get out of the chamber without decaying; no one can tell which is going to happen when. It is interesting to note that this sort of thing led

Wolfgang Pauli (Nobel Laureate 1945) to team up with Carl Gustav Jung, the famous psychologist, to write a book on, amongst other things, an acausal principle in the Universe. Jung had come to a belief in an acausal principle in a quite different way, via his study of what he called synchronicity, the too-frequent occurrence of events which are meaningfully, but not causally, connected; coincidences, for short.

Another odd thing about quantum mechanics was that it predicts only probabilities, the chances that one or another event would happen given certain initial conditions. An even more peculiar thing about this was noted by Heisenberg. He said

> The concept of a probability function does not allow a description of what happens *between* observations. Any attempt to do so leads to contradictions; this must mean that what happens is restricted to observations. This is a very strange result.

Very strange indeed. To emphasise the strangeness Erwin Schrödinger envisaged an experiment.

Figure 1.3 is a diagram of this experiment. We have a box,

Fig. 1.3. Schrödinger imagined a box big enough to hold a cat and some apparatus. The apparatus consisted of a speck of radioactive material, a geiger counter, and some electronics to amplify the counter signal and use it to open a cylinder of hydrogen cyanide. The electronics is only switched on when the lid of the box is closed.

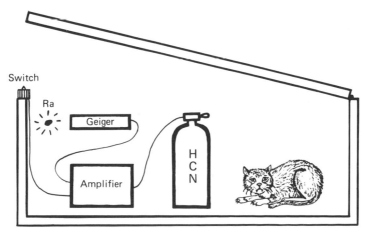

Library - St. Joseph's College
222 Clinton Avenue
Brooklyn, N. Y. 11205

big enough to hold a cat and some apparatus. The apparatus is a speck of radioactive material, a geiger counter that can detect the decay of an atom in the speck, and an amplifier to turn the tiny pulse of electricity from the geiger counter into enough current to open an electromagnetic valve on a cylinder of hydrogen cyanide. The electronics is turned on when we close the lid of the box. After that, the single 'quantum' event of a radioactive decay causes a release of HCN which will kill the cat.

We can imagine two physicists, one an old-fashioned classical physicist and the other a quantum physicist. We ask them two questions. The first is 'I have just closed the lid and shall open it again in five minutes. What shall I find, a live cat or a dead cat?' The classical physicist says he cannot answer; classical physics cannot handle radioactive decays. The quantum physicist says 'easy: just let me know the particular element you are using and how much of it you have and I'll calculate the probability of finding either a live or dead cat, in five minutes or any other time you like to mention!' So far, the quantum physicist is well ahead. But now the second question: 'I have just closed the lid of the box; what is in there now?' The classicist says 'easy, a live cat, or a dead cat.' The quantum physicist, however, says 'I don't know – quantum mechanics does not allow a description of what happens between observations. There is a wave function, of course, developing with time, but that is not in the box, but in n-dimensional space.' This is the paradox of Schrödinger's cat.

You can see that this makes the universe a very different place to what classical physicists had thought. For one thing, it makes the observer absolutely essential – if what happens is restricted to observations, then when there is no observer, nothing happens. The philosophical consequences of the success of quantum mechanics are considerable. Heisenberg wrote two books on this, *Physics and Philosophy* and *Physics and Beyond*. Schrödinger also; *Mind and Matter* and *What is Life*. Both authors pointed to the similarities between this modern physical view and Eastern mystical philosophy.

A third difficulty is that, in a way, quantum mechanics implies speeds greater than the speed of light. We imagine some simple experiment in which we shoot one electron at a screen through a thin gold foil (like G. P. Thomson's experiment). While the electron is in flight, the quantum-mechanical wave function gives us probabilities that the electron will strike any point of the screen, which we can imagine as stretching out indefinitely in all directions. Then all of a sudden the electron strikes and the probabilities for all other points instantaneously become zero. This is called 'the collapse of the wave function'. It does not quite contravene Einstein's assumption that nothing can travel faster than light because by 'nothing' he meant 'no material body and/or information'. But it is close enough to contravening it to make one uneasy (it made Einstein uneasy).

A fourth, rather surprising difficulty is that quantum mechanics is anti-atomic. Heisenberg said, talking about the atoms of Democritus,

> Perhaps this entire philosophy was wrong. Perhaps there did not exist any smallest elements which could no longer be divided.

Quantum mechanics sees the Universe not as a collection of discrete indivisible atoms but as a single connected whole. Admittedly, to make calculations we must make approximations and leave out the effects of large parts of the Universe on the particular small part we are considering. But this is only because our calculating ability is strictly limited. We need always to remember that we have made an approximation, for unfortunate consequences of making approximations have a habit of popping up in very unexpected places (especially when the habit of approximating has become almost unconscious).

Now we could solve all of these difficulties of quantum mechanics easily if it were not for the fifth difficulty. We could do it by just saying, 'this is a crazy theory; throw it out.' But the fifth difficulty is that quantum mechanics *works*. It is not only by far the most accurate theory physicists have ever had; also, it explains and *predicts* many things no other theory can handle. It has given us a complete theory of atomic and molecular spectra; a theory of the chemical bond (there was a famous early paper by Heitler and London which physicists say turned all of chemistry overnight into a branch of physics); theories of radioactive decay; theories of the atomic nucleus; an understanding of how the Sun (and other stars) produce their heat; transistors, integrated circuits, microprocessors, computers, lasers; the atomic and hydrogen bombs; and so on. So, crazy or not, it works.

And that, very briefly, is where the study of the 'electronic' part of the atom has taken us, so far. Now, what about the nuclear part?

Rutherford's experiment of scattering fast helium nuclei (from the decay of radium or some such element) on atoms of gold had established the size of the nucleus as $\sim 10^{-15}$ m. It was positively charged and contained most of the mass of the atom; most nuclei were about 4000 times more massive than their accompanying electrons. An exception to this was the hydrogen atom where the ratio was $\sim 2000:1$. Very early, Rutherford suspected that this meant there was a new sort of particle in the nuclei – a neutral particle, about as heavy as a proton. If this were so, then the helium nucleus would be made of two of these neutral particles and two protons.

More stolid, less intuitive people did not like the idea. It had what appeared to be obvious defects. Helium nuclei were very stable – they could hit a gold atom and bounce back without breaking up. But the only known force that could act between the *neutral* particle and a proton is gravity and that is amazingly weak, on the atomic scale. Moreover, Rutherford could not produce any of these neutral particles so the idea stayed an idea in Rutherford's mind. But it was a good idea.

Quite a few years earlier people had begun to notice that the atoms of an element were not all of the same mass. There are three naturally occurring radioactive series: one begins with uranium, another with thorium and the third with actinium. In each of the families the original atom goes through a whole series of radioactive decays and ends up as an atom of lead. One knows at each step just know much mass and electric charge the atom loses. If an α-particle is given off, it loses a mass about equal to the mass of four protons and it loses two positive units of charge. If it loses an electron, the mass loss is small but it loses one negative unit of charge (which is the same as gaining a positive unit). One can calculate the mass of the three lead atoms for the three different series: they turn out to be different, 206, 208 and 207 respectively. But they all have the same electric charge on their nucleus, 82 units, therefore the same number of planetary electrons and the same chemical properties – that is, they are what we call 'lead'. These three different forms of lead were given the name 'isotopes' of lead. The word comes from two Greek words meaning 'same place' because all three forms occupy the same place in the Periodic Table of the elements.

Now if we use some element other than hydrogen in a discharge tube (of the type J. J. Thomson used to discover the electron), the 'positive rays' will be positively-ionised atoms of that element. By suitable use of electric and magnetic fields we can determine their mass; such a device is called a mass spectrograph. By the early 1920s, Aston in Cambridge had examined some of the higher elements to see if they came in several atomic masses. (Aston was awarded the Nobel Prize for chemistry in 1922.) He found that indeed they did. Since then we have found that most of the elements have several stable isotopes. For instance, hydrogen itself has an isotope (called deuterium) whose nucleus is twice as massive as the proton. Tin has *ten* stable isotopes. All elements, also, have unstable isotopes, while some of them, such as uranium, do not have a stable form at all.

By the mid 1920s it was also known that the nucleus could be disrupted (though not to the reporter who wrote the story I mentioned at the beginning of this chapter). Rutherford had a laboratory assistant, John Kaye, who was still at Manchester when I was a student there. Kaye often built and carried out Rutherford's experiments. One method of detecting α-particles (that is, fast helium nuclei) was to let them hit a screen of zinc sulphide. The tiny crystal that was struck by the α-particle gave out a small flash of light (it is the same principle used in a modern TV screen – except the particles used there are energetic electrons and there are many more of them). If one had sharp eyes and worked in a dark room these minute flashes could be counted. And if gas was gradually admitted to an enclosure containing the source of α-particles and the screen, it would become clear how much gas was needed to stop the α-particles. This gave one a measure of their energy. One day Kaye was doing this and arrived at the point at which the scintillations stopped. But he then noticed a flash. Now, especially for a beginner, it is very easy in these circumstances, to 'see' flashes that are not there. But Kaye was not a beginner and trusted his ability. So he reported it to Rutherford, who did not believe him. Kaye stuck to his guns and in the end (by demonstration) carried the day. If the gas is air, very occasionally an α-particle will strike a nitrogen nucleus head on, disrupting it so that it emits a proton, which has a longer range than an α-particle. The atomic nucleus had been split. Later, the Cockcroft–Walton experiment broke new ground in producing disintegrations by artificially-accelerated particles. But the effect is the same.

This disruption of light nuclei by α-particles had other important consequences. Two German physicists, Bothe and Becker, found that if they bombarded beryllium with α-particles they got a very penetrating radiation. They thought at first it was very hard X-rays, but Curie and Joliot in France showed that these rays behaved differently to X-rays. They passed through sheets of metal almost uninfluenced, but if a sheet of paraffin wax was placed between them and an ionisation chamber the ionisation was nearly doubled. Chadwick, who had studied under Rutherford in Manchester and was then (1932) working with him in Cambridge, leapt upon this observation and quickly showed that the unknown radiation was knocking protons out of the paraffin wax and that it consisted of particles of about the same mass as the proton but without any electrical charge. This absence of charge meant the particles did not lose energy by tearing electrons from atoms they passed and hence could penetrate materials like lead much more freely than X-rays.

These particles were the particles Rutherford had hypothesized over ten years earlier. Chadwick called them neutrons. It was a most important discovery, and recognised as such. In 1935 the Nobel Prize for physics went to Chadwick, and the chemistry prize to Irene Curie and M. F. Joliot. Bothe shared the physics prize in 1954.

The neutron was important in two ways. First, it confirmed Rutherford's idea that nuclei are made up of protons and neutrons (except for ordinary hydrogen, whose nucleus is a single proton). Deuterium – heavy hydrogen – has a nucleus composed of one proton and one neutron. The usual form of helium has a nucleus made of two neutrons and two protons, although it has isotopes with one and three neutrons respectively. And so through the elements. As the atomic weight increases, the number of neutrons tends to become somewhat larger than the number of protons. The heaviest stable nucleus is that of bismuth, which contains 82 protons and 126 neutrons. So we now knew of what nuclei were made.

Secondly, the neutron was important because it makes a much better projectile for altering other nuclei than does an α-particle. This is because it is electrically neutral. The nucleus itself is positively charged, so it repels the positively-charged α-particle. The α-particle must have a considerable energy to get into the nucleus through this 'Coulomb barrier'. (Coulomb, a French physicist, was the first person to establish the law of force between electric charges.) The neutron, being neutral, does not notice this electrical barrier and slides into the nucleus unhampered by it. The discovery of the neutron made it possible to cause and observe many nuclear reactions. It made the transmutation of elements routine.

This field of study is now called 'nuclear chemistry'. At first people thought these modern physicists had achieved the dream of the old alchemists, but this has been shown, by the very detailed work of C. G. Jung and his co-workers, to be a misinterpretation of the aims of the alchemists. Repeatedly the alchemists wrote '*Aurem nostrum non est aurum vulgum*', ('Our gold is not the common gold'). What the alchemists were after was the much more difficult task of the transformation of their psyche.

So by 1932 we had found out that the ordinary matter around us is made of molecules; that the molecules are made of atoms; that the atoms consist of a comparatively large 'electron cloud' (called a cloud because the Uncertainty Principle tell us we cannot locate the electrons with perfect accuracy) and a much smaller nucleus; and that the nucleus is made of protons and neutrons. There was one obvious and outstanding problem left. The nucleus had been 'split' in quite a few cases – but it took a lot of energy to do it. Obviously the protons and neutrons are tightly bound together. The enormous density of nuclear matter showed this. But how? Gravity is much too weak a force. The electrical force between a proton and neutron is zero because the neutron has no charge. What could be doing it? It took four years to come up with a reasonable theoretical answer and *fifteen* years to establish that answer by experiment.

Before this happened, there was another development of significance. The original quantum-mechanical treatments of the hydrogen atom had been non-relativistic. With heavier atoms one deals with greater forces and, in a classical sense, faster moving particles. To deal with these one needs to incorporate the theory of relativity. P. A. M. Dirac, of Cambridge, was able to do this and start what is now called quantum electrodynamics (often called QED for short).

His theory had what appeared at first sight to be a considerable defect. It predicted that the electron should have a positively-charged counterpart – what we would now call an anti-particle. No such particle was known at the time. The proton, which had the right charge, had very much the wrong mass – it is about 2000 times heavier than the electron. Then, in 1932, a young American physicist, Carl Anderson, found the positive electron in cosmic radiation. So I had better backtrack again and explain a little about cosmic radiation. Sometimes in the last few years, I have been to places like the Esalen Institute near Big Sur, California, or the International Transpersonal Conferences. People in these places sometimes ask me what I do. I quickly found that if I say 'physics', their eyes get a kind of glazed look and they quickly wander away. But if I say 'I study cosmic radiation', their more likely response is 'wow man! way out!', and closer contact becomes possible.

Cosmic radiation was discovered in 1912 by an Austrian, Victor Hess (he shared the Nobel Prize for physics in 1936 with Carl Anderson). The discovery arose out of experiments, some of which dated back to the early days of radioactivity. The radiations from radioactive substances (the α, β and γ radiations) all ionise gases when they pass through them. In fact they knock electrons out of atoms in their path, leaving behind them a trail of separated charges, half positive and half negative. This makes it possible for the gas to conduct electricity more easily. We can detect these ions, as they are called, in various ways. Geiger counters do it. One very simple method is the ionisation chamber in which we have two electrodes in the gas (see fig. 1.4). We arrange to have one electrode positively-charged and the other negatively-charged and we measure the electric current that flows between them when the gas between them becomes ionised. With these ionisation chambers people found that in any laboratory there was always *some* radiation passing through the ionisation chambers. In a uranium mine there was a lot. It seemed likely that some of the radiation in other places was due to the small amounts of uranium and thorium in any dust.

But people took ionisation chambers out to sea, where there is not so much dust, and still found radiation. Then Wulf tried taking an ionisation chamber up to the top of the Eiffel Tower in Paris and found more radiation than he expected. So Göckel tried balloon flights with a similar result. In 1912, Hess made eight flights, the seventh of which reached an altitude of 5350 m. He found that the ionisation decreased up to about 800 m and then began to increase again. It reached the ground level value between 1400 and 2500 m. At 4000 m it was six times the ground level value and above that increased very rapidly. Hess drew the correct conclusion, namely that the Earth was being bombarded by a very penetrating radiation.

The study of the radiation was brought to a halt by World

War I. It got going again rather slowly after 1919, and one of the main schools in the early days was Kohlhörster's in Berlin. New instruments were brought into play: the original geiger counter was improved to become the Geiger–Müller counter; C. T. R. Wilson's cloud chamber was brought into a vertical orientation by the Russian, Skobeltsyn, and coupled with the G–M counter by both Blackett in Cambridge and Anderson at Caltech, Pasadena, to become the counter-controlled cloud chamber. By 1930 a young Dutch physicist, Clay, had taken an ionisation chamber on a long sea voyage (a good way to do physics) from Holland to what was then the Dutch East Indies and shown that, because the Earth's magnetic field reduced the

Fig. 1.4. A sketch of an ionisation chamber. A cosmic ray particle passing through separates electrons from previously neutral atoms, leaving a wake of positive and negative charges behind it. The two electrodes, one positive, the other negative, collect these charges by electrostatic attraction and the resulting electrical pulse is then amplified electronically.

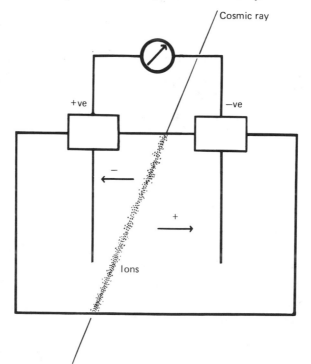

amount of radiation, the radiation incident on the Earth must be positively charged. It turned out to be a very mixed bag – but mostly protons. These particles were *very* energetic, especially for those days when 10 million electron volts seemed an enormous energy. Some of the cosmic ray particles were at least 1000 times as energetic as that. (No one has yet found a limit to the energy they can have; probably we, in Sydney, hold the present world record with an event detected by our giant air shower array, operating in the Pilliga State Forest near Narrabri. 'Giant' is quite large – a collecting area of 60 km². The event had an energy around 10^{22} eV!) These energetic particles collide with air nuclei near the top of the atmosphere and produce other particles, which in turn produce others and the cascade spreads down to sea level.

Carl Anderson was studying this radiation with a Wilson cloud chamber in a magnetic field. The chamber was quite small (the method is described in detail in chapter 3), 16.5 cm in diameter and 4 cm deep. The magnetic field was parallel to the axis of the chamber. Once the chamber is expanded, small drops of moisture form on any ions present and so the path of the charged particle that went through the G–M counter is made visible. Because it is a charged particle it interacts with the magnetic field and is deviated from its usual straight trajectory. In a constant magnetic field, perpendicular to its direction of motion, its path is part of a circle. The radius r of the circle is determined by its charge ze, the strength of the field H, and its momentum p. The bigger the charge, or the field, the smaller the radius of the circle; the bigger the momentum, the larger the radius of the circle. Briefly, $r = Ap/Hze$, where A is a constant depending on what units we use to measure the various quantities.

Figure 1.5 shows a diagram of the event that proved the existence of the positron. The horizontal bar across the middle is a lead plate, 0.6 cm thick. The track of the positron moves upwards from lower left, through the plate, to the upper left. The direction of motion is given by the curvatures below and above the plate. Obviously the radius of curvature is smaller above the plate than below it. Because it can only lose momentum in forcing its way through the plate, this gives the direction of motion. It is a bit unusual to find cosmic ray particles moving upwards. That does not mean that they have come through the Earth; it means they, or their parent particles, have been back-scattered from the Earth or materials in the laboratory.

Once we know the direction of motion and the direction of the magnetic field we know the sign of the electric charge on the particle. For this particle it is positive. Then from the ionisation produced per unit length of its track and the loss of momentum in penetrating 0.6 cm of lead, we can calculate both the magnitude of its charge and its mass. For this particle, the charge was of the same magnitude as the charge on an electron and its mass was close to that of an electron – certainly not anywhere near the 2000 times greater mass of a proton.

Fig. 1.5. A diagram of Anderson's cloud chamber photograph that established the existence of the positron. A strong magnetic field perpendicular to the plane of the paper curved the track of the positron. The lower the energy the greater the curvature.

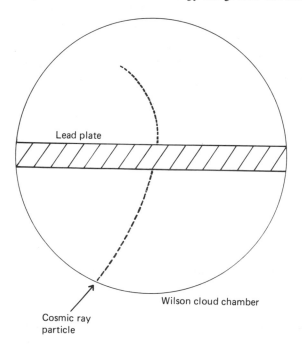

Lead plate

Wilson cloud chamber

Cosmic ray particle

Anderson then concluded he had found a positive electron and it was the particle predicted by Dirac. His result was very quickly confirmed by Blackett in England and by others. The first of the anti-particles had been discovered, the first example of what we now call anti-matter.

The way in which these positrons arose was soon discovered. They are not there all the time, like electrons. Our present Universe is a very dangerous place for positrons. There are many, many electrons around and when a positron meets an electron they are both annihilated – their energy appears as γ-radiation. The positrons are produced when a sufficiently energetic γ-ray passes close to an atomic nucleus. Then the γ-ray may lose its energy and produce a pair – one positron, one normal electron. Positrons are never produced alone.

So by 1932, the neutron had been discovered, we had some idea of the constitution of the atomic nucleus, knew at least one anti-particle and realised the earth is bathed in a very high energy charged radiation coming from somewhere well outside the Solar System. The big problem remained: what holds the protons and neutrons so tightly together in the atomic nucleus?

The solution to this came from Japan. (It is interesting to note just how international this quest is.) It was put forward by Dr Hideki Yukawa of Kyoto. Let us look at the difficulties that faced him. He needed a very *strong* force, for the particles in the nucleus are very tightly bound to each other. He needed a force that operated between neutral and charged particles as well as between charged particles. He needed a force with only a very short range. This was quite different to the forces of electromagnetism and gravity. Both of these have infinite range (or as close to infinite as we have been able to check). Gravity pulls an apple down to earth; it keeps the Earth moving round the Sun and the Sun moving (along with a thousand million other stars) round the centre of mass of our galaxy. It keeps our local cluster of galaxies together. It is a long range force. So, as far as we can tell, is the force between two electric charges, or two magnetic poles. But this new force cannot have an infinite range – otherwise it would be so strong it would pull the whole

Universe together into a superdense glob. From the standpoint of 1932, this was a very odd force.

Yukawa solved the problem by looking at the electromagnetic force and altering it to suit. The electromagnetic force is 'carried' by photons. Photons are the particles of light. They always move at the speed of light. Having no rest mass, they can never be at rest! Their energy (and hence their mass) is hv where h is Planck's constant and v is their frequency (looking at them as waves, this time). Two electric charges exert a force on each other by exchanging virtual photons. 'Virtual' in this context means very short lived. The length of time a virtual particle can exist is given by the uncertainty relation, $\Delta E \Delta t \geqslant h$. Since photons have no rest mass we can make the uncertainty in mass/energy, ΔE, as small as we like. So we can make Δt as large as we like. The distance they can travel is just $c\Delta t$, where c is the velocity of light, so the range of the force is infinite.

All Yukawa required to get a short range force was to suppose the new force was 'carried' by a particle with a finite rest mass. Moreover, since he knew from Rutherford's scattering experiment just what the range was ($\sim 10^{-15}$ m) he could estimate the mass of this 'quantum'. It came out as ~ 300 times the rest mass of the electron – in between the mass of the electron and that of the proton. For this reason, the new particle was called a 'meson', meaning 'middle one'.

For the protons and neutrons to attract each other, he supposed, they both carried the same 'nuclear charge' and, in this field, like particles attract each other. The detailed theory shows this depends on the spin of the particles – its intrinsic angular momentum. The photon has a spin of $+1$ and like electric charges repel each other. Yukawa supposed his meson to have a spin of zero. Finally, he supposed the 'nuclear charges' were basically stronger than the electric charge.

He was able to predict some other features of his particle. Because the particle was quite heavy, that is, it had quite a lot of rest energy, it should be unstable – radioactive if it were given the chance. Normally, in binding together a neutron and a proton, the meson was just emitted by one and, very shortly afterwards, absorbed by the other. But if somehow it was given enough energy to become 'real' rather than 'virtual' and move away from its parent proton, then it should only live a fairly short time (around 10^{-8} s) before decaying, probably to an electron or positron (if it were charged) and a neutrino. Finally, Yukawa looked around for some situation where there *was* enough energy and realised that the collisions cosmic rays made in the atmosphere was the most likely situation. So he predicted that the particle might be present in cosmic radiation.

Within a year Carl Anderson and Seth Neddermeyer found some evidence for the existence of such a particle in cosmic radiation and, by 1938, using their magnetic cloud chamber at high altitude (4300 m), they obtained a photograph of a meson stopping in the cloud chamber gas. They estimated its mass as ~ 240 times the mass of the electron, noted that its electric charge was positive and thought there might be the track of a positron emerging from the end point of the meson track – meaning that the meson had decayed to a positron plus one or more neutral particles. It looked as though Yukawa was justified.

However, when people began to investigate this cosmic ray meson in detail they found some puzzling differences between its properties and those Yukawa had predicted for the 'nuclear force' meson. Some of these seemed rather niggling. Its mass turned out to be close to 200 electron masses, instead of close to 300 as Yukawa had estimated. Its lifetime was around 2×10^{-6} s instead of $\sim 10^{-8}$ s. Now these are both short times by our standards, but one *is* 200 times longer than the other, quite a difference. But the really big difficulty was that whereas Yukawa's particle was the particle that bound the proton and neutron together and, necessarily, had a very strong interaction with them, the cosmic ray meson showed an amazing ability to penetrate matter. In other words, *not* to interact with it, other than by the electromagnetic ionisation process. People found that a 10 cm thick wall of lead made almost no change in the rate of such particles. A London physicist took counters down into a Tube station and still found plenty of mesons. Then a coal mine was tried and the mesons, somewhat reduced in intensity, still came through the roof. It was very puzzling. At this very interesting point, World War II broke out and research into the cosmic radiation was much reduced.

After the war, big efforts were mounted to solve the problem. The 'cosmic ray' mesons are quite plentiful, even at sea level, and some quite big experiments were set up to investigate their properties in detail. But, as often happens, it was a small experiment, apparently out of the mainstream of cosmic ray research, that solved the problem.

Radioactivity was discovered by Bequerel in 1896 using photographic emulsion as the detector. But from that point on, until 1945, photographic emulsion was hardly used. Emulsion is a suspension of small grains (0.1 to 1 μm in diameter) of silver bromide in gelatine. It was found to be difficult to get enough sensitised grains from the ionisation produced by a singly-charged particle to get a recognisable track. Moreover, the grains are very small so a microscope is needed; magnifications of 2000 times are often used. Also, normal emulsion thicknesses were very small – 10 to 20 μm – so that unless a particle was fortunate its trajectory soon left the pellicle.

Between the wars, the sensitivity of emulsion was improved somewhat, but what was really needed was a greater concentration of silver bromide – more bromide, less gelatine. In 1945, this was achieved. Emulsions containing about eight times as much light-sensitive silver bromide as ordinary emulsion were made by Demers in Canada, and by Dodd and Waller of the Ilford Company in England. Cecil Powell and his group had been working with photographic emulsions throughout the war (in collaboration with Ilford). They exposed some of the new emulsions to cosmic radiation at 5500 m altitude in the Bolivian Andes (one of the group, C. M. G. Lattes, is a Brazilian). At such a high altitude, the cosmic radiation is much more intense than at sea level. These plates, when processed, showed the tracks of particles that could be identified by their ionisation and range as intermediate in mass between protons and electrons.

The group, and others, had seen such tracks in previous

exposures. In this new stack the number of stopping mesons was so great they could identify different types of behaviour. One type, which they labelled π-mesons, stopped and a single track appeared emerging from the stopping point. This track was also that of some sort of meson (that is, its mass was intermediate between that of an electron and that of a proton). They found eleven cases where this second track also stopped in the emulsion. In all cases the length of this secondary meson track was the same. This, coupled with the similar density of grains along the track, means that all 11 had similar masses and all had started out with similar momenta. This makes it look very much as if a heavier meson had stopped in the emulsion and decayed to a lighter meson (and some other particle that did not leave a track in their emulsion, either because it was uncharged, like a neutron, photon or neutrino, or because it was too lightly ionising, like a sufficiently fast electron).

In other cases the stopped meson gave rise to what they called a 'star'. This is a number of tracks coming from one point and is due to a nucleus of the emulsion being disintegrated. The common nuclei in emulsion are silver and bromine (from the silver bromide) and carbon, nitrogen, oxygen, and hydrogen (from the gelatine).

Both of these types of event were sometimes caused by mesons that themselves came from 'stars', that is, were observed to come from nuclear disintegrations. Then there were mesons that came into the stack from outside and just stopped, seeming to do nothing (but they may, of course, have decayed to fast, light, charged particles, which the emulsion was not sensitive enough to record).

The Bristol group knew about the previous cosmic ray work – that mesons, of one sort at least, existed, had a mass around two hundred times the electron mass, a lifetime around 2×10^{-6} s, and were very penetrating. And they knew of Yukawa's prediction. They saw this welter of facts could be explained if Yukawa's theory was altered a little. They supposed the Yukawa meson is produced in high energy nuclear encounters in both positively-charged and negatively-charged forms, and has a life-time (as Yukawa had suggested) around 2×10^{-8} s. If a positive π-meson (as they called it) stops in emulsion it cannot enter a nucleus because it is repelled by the positive electric charge on the nucleus. It wanders around the atomic electron cloud till it decays – to a μ-meson and one other particle, probably a neutrino. They could tell that the pion (the modern name for a π-meson) decays to only two particles because of the unique range of the muon. If the pion had decayed to three or more particles the total momentum could be shared amongst them in various ways, so that the muon would not have a unique range.

On the other hand, if the pion is negative it is attracted into the nearest nucleus. Because it is the Yukawa meson, it interacts violently with this and disintegrates it. It may be quite slow, but its rest energy is about 150 MeV. This disintegration shows up in the emulsion as a 'star'.

Finally, the mesons that came into the plate and appeared to do nothing were muons that had been produced via pions outside the plate. Because their life-time is about one hundred times longer than that of pions, they are about one hundred times more plentiful in the atmosphere.

There were a few little points to tidy up. The pion comes not only in positive and negative forms but also in a neutral form – so there are three varieties of pion, positive, negative and neutral. The neutral pion can decay in an even shorter time than 10^{-8} s, namely $\sim 10^{-16}$ s. It decays in a somewhat different way, into two high energy photons, two γ-rays. This is forbidden to the positive and negative pions because of the law of conservation of electric charge. So we have

$$\pi^{\pm} \rightarrow \mu^{\pm} + \nu,$$

$$\pi^0 \rightarrow 2\gamma.$$

The muon decays to an electron and *two* neutrinos. This decay had been seen in Wilson cloud chambers before World War II. The electron was too fast and lightly ionising to be seen in the early nucleus emulsions. By 1950, Kodak had produced an 'electron sensitive emulsion' and the whole decay sequence could be seen in the emulsion,

$$\pi^+ \to \mu^+ + \nu,$$
$$\mu^+ \to e^+ + \nu + \nu.$$

By mid-1947 then, the atomic hypothesis seemed in very good shape. We knew that matter is made of molecules; molecules are made of atoms; atoms consist of an electron cloud about 10^{-10} m across and a tiny, heavy nucleus about 10^{-15} m across; and the nucleus consists of protons and neutrons bound together by pions. It looked very elegant and complete. There were a few oddities – why the muon? It seemed an unnecessary particle. Why not

$$\pi^\pm \to e^\pm + \nu,$$

why the odd, short-lived extra? Then the positron. If the positive electron exists (and it does) why not a negative proton? And then again, quantum mechanics seemed to the experts to be unhappy with the atomic hypothesis as its running mate. But, in mid-1947, one could easily ignore these difficulties – dismiss them, as just a few t's to cross. In the 18th century, Alexander Pope must have felt the same way; he wrote

Nature and Nature's Laws lay hid in night.
God said 'Let Newton be', and all was light.

In the early 20th century Sir John Squire added an extra couplet:

It did not last. The Devil howling 'Ho!
Let Einstein be' restored the status quo.

Something similar was about to happen in particle physics, but this time it was not Einstein who was to upset the apple cart. It was two young physicists at the University of Manchester, George Rochester and Clifford Butler.

2

The strange particles

In December 1947 I was invited to a small conference in the Dublin Institute of Advanced Studies. The Dublin Institute had been founded by the Irish government before World War II. The Irish saw the plight of Jewish and liberal scientists in Central Europe at that time. Eamon de Valera, the Taoiseach, a mathematician, saw he could help some of the scientists, and his own country, at the same time. So the Institute was started, with two schools, one for Theoretical Physics and one for Celtic Studies. Both have been very successful.

The Theoretical Physics school was established around two of the leaders of the early days of quantum mechanics, Erwin Schrödinger and Walter Heitler. After World War II de Valera decided to extend his very successful Institute by the addition of a School of Cosmic Physics, comprising three sections, meteorology, astronomy and cosmic radiation. He asked Lajos Janossy, another refugee who had spent World War II working in P. M. S. Blackett's laboratory in Manchester, to take charge of the cosmic ray section. Janossy had been one of my supervisors when I took my Master's degree at Manchester and he asked me would I be interested in the Assistant Professorship. So I went to Dublin to be inspected (and to inspect) as well as to attend the small conference.

One of the leading speakers at the conference was George Rochester, my other supervisor. He showed two Wilson cloud chamber pictures that he and Clifford Butler had taken with Blackett's magnetic cloud chamber. I knew that apparatus quite well and, when I finished my Master's degree, got it running again (under Janossy's and Rochester's direction) to do the experiment being reported by Rochester. But World War II prevented us getting very far. However, after the war, Rochester went back to that experiment and these pictures were the result. They are shown in figs. 2.1 and 2.2.

To see what Rochester and Butler were after in this experiment we need to backtrack a little. Very soon after Anderson discovered the positron in 1932 he (in Pasadena), Blackett (in Cambridge), and others made a further discovery. A cloud chamber, operated with a lead plate (or maybe several lead plates) in the middle, often revealed the track of a single particle entering the cloud chamber, hitting the lead plate and then a shower of particles emerging from the lead plate. In this shower

Fig. 2.1. The discovery of the V^0 particle (now called a K^0 meson) by George Rochester and Clifford Butler. The Wilson cloud chamber picture shows many mesons entering the cloud chamber from a high energy interaction in the lead above the chamber. Many of these particles (which are probably pions) penetrate the thick lead plate in the middle of the chamber. The tracks of interest form an inverted V just below the lead plate and to the right of centre.

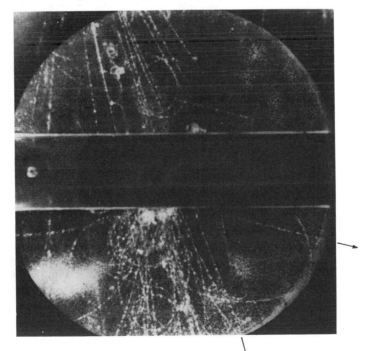

about half the particles were positrons and half were 'ordinary' negative electrons. The processes leading to the generation of this shower were quickly explained by the new generation of quantum-mechanical theorists – Heitler, Bethe, Bhabha, Oppenheimer, and others. If the initial particle is an energetic electron and if it passes close to a nucleus of lead, it is deflected by the large positive charge on the nucleus. That is to say, it is *accelerated* and, being an electric charge, when accelerated it emits electromagnetic radiation. If the collision is close enough the radiation is an energetic γ-ray. After the collision we have a somewhat less energetic electron plus an energetic γ-ray. The

Fig. 2.2. The discovery of the charged V particle, also by Rochester and Butler. The track enters the chamber at top right, deflects sharply after a few centimetres and then penetrates the thick lead plate without interaction. The probable decay is $K^+ \rightarrow \mu^+ + \nu$. This, and the preceding photograph were generously provided by Professor G. D. Rochester FRS.

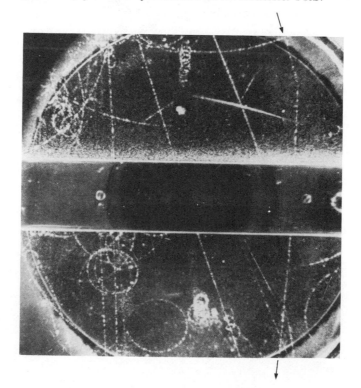

electron can carry on doing this until it has lost all its energy. But the γ-ray can also interact. According to the then new quantum electrodynamics (QED), if the γ-ray passes close to a nucleus it creates an electron–positron pair (and, itself, disappears). Then both the electron and the positron can carry on the process (until their energy becomes too low) and a shower develops.

Lead is very good for promoting this process because of the large electric charge on its nucleus. For much the same reason it is good at absorbing the shower. Lead 10 cm thick will stop almost any electron. But a group of physicists in Brazil, led by a Russian emigré, Gleb Wataghin, found that some showers could penetrate much greater thicknesses of lead. The first sort of shower was called 'soft', the second type 'penetrating'. Scientists suspected showers of the second type were not electromagnetic but were in some way connected with the strong nuclear force and the Yukawa meson. It was a hot topic in 1940 and my Master's thesis was on an investigation of these showers using a Wilson cloud chamber (but no magnetic field). Rochester and Butler were engaged on a similar experiment but this time with a magnetic field.

Figure 2.3 shows the experimental set-up Rochester and Butler used. It was designed to record those events when an energetic cosmic ray particle collided with a lead nucleus and produced a penetrating shower. Some of the particles belonging to this shower would go through both top and bottom lead blocks, discharging the buried geiger counters, triggering the cloud chamber to expand and photograph the tracks within it. The cloud chamber had a thick (3 cm) lead plate across its mid plane. So if any of the particles were electrons or positrons they would produce a soft shower. If not, then they might be Yukawa particles (this experiment was designed and began operation before the discovery of the pion).

What Rochester and Butler discovered was even more surprising than the pion. The pion, after all, was a predicted particle. The two particles in figs. 2.1 and 2.2 were completely unexpected by theory. Consider fig. 2.1. It shows the sort of event that the experiment was set up to detect. A shower of particles emerge from a point in the lead block above the chamber. Some of them penetrate the 3 cm lead plate in the chamber without producing a soft shower, and so are not electrons. (It is very likely that they are pions.) But the interesting event is the pair of tracks making an inverted V just below the lead plate and right of centre.

Fig. 2.3. A diagram of the Rochester–Butler experiment. The cloud chamber was triggered by coincidental discharges of the Geiger–Muller counter in the three banks of counters above and below the cloud chamber. Some of the counters were buried in thick lead so this could only happen when the cloud chamber was struck by an energetic cosmic ray shower.

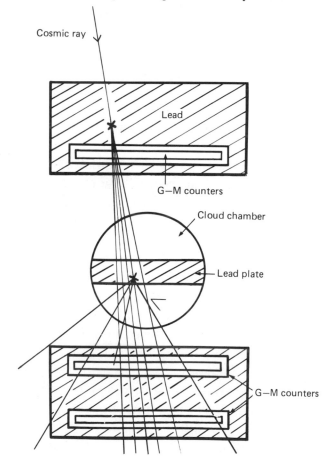

At first thought, these might not be the tracks of two separate particles; they could be the track of one particle, either moving upwards and scattering through more than 90 ° to the right or coming in from the right (moving somewhat upwards) and scattering downwards. However there are several objections to this explanation, one of which is overwhelming. The cloud chamber has a strong (3500 gauss) magnetic field along its axis, so the trajectories of all charged particles are circles. The radius of the circle is a measure of the momentum of the particle. At first glance, those two tracks look straight, indicating they are of high momentum. So if this is one particle scattering on the nucleus of an atom of gas in the chamber, then that nucleus, having suffered such a considerable recoil, would leave a noticeable track in the chamber. But it is not there. So this is not a scatter.

The only reasonable interpretation is the decay in flight of a neutral particle, produced in the original interaction, in the lead above the chamber or in a secondary interaction in the 3 cm lead plate in the chamber. Such a decay must conserve electric charge so one of the two tracks forming the V should be due to a positively-charged particle and the other to a negatively-charged particle. They do indeed show opposite curvatures when measured with precision. Their momenta were about 300 MeV/c each – very energetic particles. From this we can calculate the mass of the original neutral particle, assuming the two decay products are various possible combinations of the then known particles. The minimum possible mass is 770 ± 200 times that of an electron. This neutral particle is a lot heavier than either the muon or the pion. It could not be a neutron for at least two good reasons, one being that a neutron's half-life is ~ 12 minutes and this particle had decayed after about 10^{-9} s. Rochester and Butler had here what looked like a new and unpredicted particle. Naturally, they wanted another example of the same thing.

What they got is shown in fig. 2.2. The interesting track is the right-hand one in the picture. It too is associated with other tracks which form a penetrating shower but it is not the track of a neutral particle decaying to two charged particles. Rather, it is the track of a *charged* particle decaying to another charged particle, plus one or more neutrals (which leave no tracks). The same argument as before shows that it cannot have been a scatter and that its minimum mass is 490 ± 60 times the mass of an electron – that is, its mass was much greater than any known radioactive charged particle. (A year earlier the French physicists Leprince-Ringuet and l'Heritier had found a track of a particle in a cloud chamber which appeared to have a mass around 990 electron masses but was not seen to decay.)

Rochester and Butler it seemed, in about one year's work at sea level, had found two new and unpredicted 'elementary' particles. And, indeed, it turned out to be so. But confirmation was slow. The obvious way to get more of these events was to move the apparatus to a mountain – about 3000 m high, or higher. At that altitude the cosmic radiation is 30 or 40 times more intense than at sea level, and the rate of events that much greater too. However, such mountains do not exist in England and the Manchester group stayed at sea level. But Anderson, in Caltech, was used to work at high altitudes and his group began to operate at 3350 m. In 1950 they published a paper giving details of 30 of the 'neutral V' events (like Rochester and Butler's first event) and 4 of 'charged V' events. Mountains suddenly became very popular with cosmic ray physicists (including the Manchester group).

A similar scramble for height began amongst groups using photographic emulsion. However, a suitable stack of photographic emulsion was an intrinsically lighter and simpler device than a Wilson cloud chamber, particularly a Wilson cloud chamber in a strong magnetic field. Nor did emulsion stacks require any electrical supply. So emulsion stacks could be flown to very great altitudes using balloons. When they had had a sufficient exposure, the package could be released by radio command and descend by parachute.

The method had its difficulties. A member of my group, working with the Bristol group, flew a stack from Southern Italy in 1960. It was supposed to ascend to about 30000 m, catch an easterly air stream, drift over Italy out into the Mediterranean, and there be picked up by NATO warships. However, the balloon developed a slight leak and only got to 18000 m. It then slipped into a westerly airstream and began to move towards Albania. The Italian Air Force were asked if they could shoot the balloon down into the sea because our radio command was out in the Mediterranean – not the Adriatic. It proved to be too high for them and the balloon continued over the Adriatic and disappeared completely.

Generally however, things went better than that. At the same time, I was flying an even larger stack out of Alamagordo, New Mexico, with the USAF. This was a near–perfect flight: balloon release at dawn, a quick ascent to 31 500 m, a six hour flight to the edge of the balloon range, and a descent almost onto the heads of the recovery crew. Then a month in a dark room in Chicago processing the 20 l of emulsion and, in the process, pouring about 30 kg of silver down the sink.

The cosmic radiation is so intense at these great altitudes that six hours gives quite a good exposure. This is greatly helped by the emulsion being continuously sensitive. A Wilson cloud chamber, on the other hand, is only sensitive for a very brief period, about 0.02 s shortly after expansion. So Wilson cloud chambers need some form of triggering device that tells them when to expand, selects events promising to be 'interesting' and then expands the chamber and photographs the tracks. This is another reason why cloud chambers are best used from a terrestrial base.

It turned out, also because of their technical differences, that cloud chambers were better looking at the 'neutral V' events and emulsions were better looking at new types of charged particles. In many ways the two techniques were complementary. From 1950 to 1955 most of the progress in the study of what had been called elementary particles was made using these techniques and the cosmic radiation. There was an interesting change in nomenclature during that time. The particles became, first, 'fundamental' rather than 'elementary', and then 'strange', for reasons we shall see.

The investigation of the new particles turned out to be a tricky business, with apparent contradictions and surprises popping up all over the place. Part of this was due to the imprecisions of measurement, and much to having to use cosmic radiation as the source of energetic particles. Up till 1955, accelerators were just not energetic enough to cope with any particle heavier than a pion. Now, accelerators produce nice, tidy, intense beams of particles (when they are big enough). Today, with an accelerator one can specify in advance what *sort* of particle is needed as the projectile – protons, anti-protons, pions, both positive and negative, muons, neutrinos, neutrons, and so on down a long list. The energy of the particle can be specified, also its direction of arrival, its time of arrival and its *place* of arrival. We can arrange to have a large and precise collection of detectors waiting for it. Almost always the *number* of particles available, that is, the *intensity* of the beam, is very much greater than in cosmic radiation. With cosmic radiation it is different. One does not know when the next particle will arrive, what it will be, what its energy will be, or its direction. So precise measurement of particle masses, lifetimes and so on are hampered.

Having said all this, we can turn to an event found in 1949 by the Bristol group, which appeared to be a new particle and for which the mass could be determined with considerable accuracy. The event showed a meson stopping in a photographic emulsion. From its end point came the tracks of three charged particles (fig. 2.4). One of the particles, *a*, stopped after quite a short distance and produced a nuclear disintegration characteristic of a negative pion. Particle *b* also seemed to be a pion since its mass was determined as (285 ± 30) times the electron mass m_e. The mass of a charged pion is $273.3\ m_e$. It seemed likely that the third particle, whose mass could not be found so accurately, was also a pion. The event also had the remarkable feature that all three pion tracks lay in the one plane. This means no neutral particle can have been emitted other than in that same plane. If it had been, then its component of momentum perpendicular to the plane would have had to be balanced by the charged particles – which then would not have been coplanar.

The chance of getting four particles emitted in a plane is very small. So the original particle had decayed to three charged pions, *and three only*. Its mass was converted into the rest mass of the three pions, plus their kinetic energy. By this time the

Fig. 2.4. The first observed decay of a τ meson (now called a charged kaon). The decay is $K^{\pm} \rightarrow \pi^{\pm} + \pi^{\pm} + \pi^{\mp}$. This event was found by the Bristol group in a stack of photographic emulsion exposed to cosmic radiation. I am indebted to Professor P. F. Fowler FRS for the photograph.

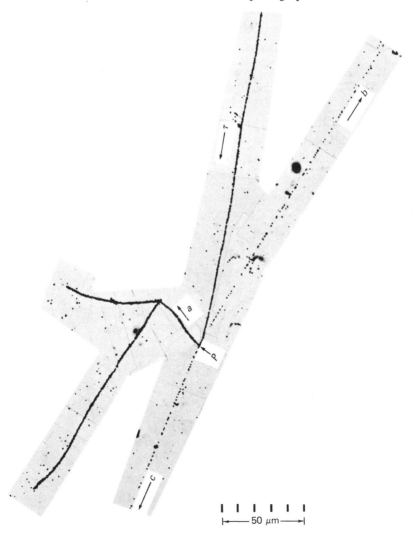

50 μm

mass of the charged pion had been accurately determined in accelerators and good estimates of the energies of the three pions could be made. The mass of the new particle followed; 969 times the electron mass. This was a particle, much heavier than a pion, and about half as massive as a proton. The 'charged V' particle Rochester and Butler found had, very roughly, the same mass – but it had decayed to only one charged particle. The particle found by Leprince-Ringuet and l'Heritier also had a similar mass – but no decay was detected.

By the end of 1952, other members of the Bristol group had found a charged meson, of mass around 1300 m_e, which decayed to a muon, and another of mass around 1100 m_e which decayed to a pion. This picture grew more and more complicated and puzzling. It was resolved in 1955 – and not by cosmic ray physicists. By 1955, the Lawrence Radiation Laboratory at Berkeley, California, had activated their new large accelerator. It was called the Bevatron. It could accelerate protons to ~ 5 GeV, that is, 5×10^9 eV. It was able to produce a *beam* of charged mesons of mass ~ 970 m_e and fire these into photographic emulsion. Because these particles had practically the same energy, they all stopped after traversing almost the same amount of emulsion – and proceeded to decay in *six* different ways. Instead of having six different particles we had one particle with six different modes of decay! This particle became known as the charged kaon; the 'charged V' particle of Rochester and Butler, the τ meson, the other two mesons from the Bristol group are all examples of one or other decay mode of this particle.

Meanwhile the cloud chamber people had had their difficulties as well as triumphs. The 'neutral V' events had an opposite idiosyncrasy to the charged kaons – two different particles could produce rather similar looking events. The first few events looked like fig. 2.1 – a V with two arms that looked similar. But then events began to turn up in which one arm was definitely thicker, more heavily ionising than the other. It became obvious this arm was the track of a *proton*. Since a proton was one of its decay products, the original neutral particle must be *heavier* than a proton! This was very unexpected. The particle was called the Λ (lambda) particle, often written Λ^0 to emphasize the fact that it is neutral – it has zero electric charge.

But then as better mass values were obtained it became clear that some of the neutral particles producing the V tracks were not as heavy as protons. They turned out to be K^0 mesons – the neutral counterpart of the charged kaons. There were two of them, the neutral kaon and the neutral anti-kaon, particle and anti-particle. The positively and negatively-charged kaons are particle and anti-particle.

The picture then became even more complicated. The Manchester group found particles decaying to Λ^0 particles. Because they decayed to Λ^0 particles, they must be more massive than Λ^0 particles. This class of particle, heavier than protons and neutrons, became known as hyperons. They are all unstable and their decay 'chain' always ends up with a proton – creating a series of decays like this

$$\Xi^0 \rightarrow \pi^0 + \Lambda^0 \rightarrow p^+ + \pi^- \rightarrow \nu + \mu^- \rightarrow e^- + \nu + \nu.$$

(These decay equations are like chemical equations. The arrow means 'turns into'. So the first equation here reads 'a Ξ^0 particle turns into a neutral pion and a Λ^0 particle'.)

With more events showing these new particles it became possible to estimate their lifetimes. For the changed kaons the mean lifetime was 1.2×10^{-8} s – rather close to that of charged pions. For the various hyperons the mean lifetimes are around 10^{-10} s. Neutral kaons behave in a very odd fashion – I shall talk about them later.

Now, these lifetimes, 10^{-8} to 10^{-10} s, sound short to us. But they are very *long* compared to the production time of the particles. This production time is easy to estimate. We imagine two protons colliding at high energy. One of them starts out at rest in the laboratory frame of reference. We can call it the target proton. It has a radius of $\sim 10^{-15}$ m and outside that distance its strong nuclear force is negligible. The other proton, call it the projectile, is moving very quickly in the laboratory system – practically at the speed of light 3×10^8 m/s. Its strong nuclear force only extends for about 10^{-15} m around it. So the two protons are only in 'strong' contact for a time of $2 \times 10^{-15}/3 \times 10^8 \approx 10^{-23}$ s. It is in this very short time that the production of the new particles must take place.

Again, to simplify, suppose the projectile is a pion and a Λ^0 hyperon is produced, that is $\pi + p \rightarrow \Lambda^0$ (10^{-23} s). Now the usual decay of a Λ^0 particle is just the reverse of this, namely $\Lambda^0 \rightarrow \pi + p$ – but this takes 10^{-10} s, some 10^{13} times longer – very difficult to understand (in 1951). One would expect the reaction rate to be about the same in either direction.

The answer was provided by A. Pais. He suggested the type of production reaction I have written, $\pi + p \rightarrow \Lambda^0$, never happens, that it is 'forbidden'. (It is interesting to ask 'by whom?'; I may touch on that later.) The new production reactions he suggested are of the form $\pi + p \rightarrow \Lambda^0 + K^0$, what he called 'associated production'. These reactions could happen in 10^{-23} s, but once the Λ^0 was on its own it could not perform the reverse reaction without the unlikely arrival of a K^0 within 10^{-15} m. By itself the Λ^0 could not decay quickly via the strong nuclear force, the Yukawa force, but only via the weak nuclear force that accounts for β-decay.

So far, I have not said much about this last force. Remember that there are three main types of radioactive decay. The first is α-decay; the radioactive nucleus emits an α-particle (which is a helium nucleus, two protons plus two neutrons). This type of decay was nicely explained by George Gamow, using the then new quantum mechanics. He imagined an α-particle trapped inside a 'potential well'. The well is generated by the combined effect of the electrostatic force between the protons and the strong nuclear force which, over a short range, tends to hold protons and neutrons together. The well is shown schematically in fig. 2.5. If an α-particle, of a certain energy E, is outside the well then it cannot get in over the 'wall' of the potential well. In other words, the positively-charged α-particle is repelled by a positively-charged nucleus. If two of the neutrons and two of the protons of the radioactive nucleus have joined up (maybe temporarily) to form an α-particle of energy E inside the well, then similarly it cannot get over the potential barrier to get out.

At least, according to classical physics it cannot. But quantum mechanics says differently. According to quantum mechanics, the α-particle always has a certain probability of 'tunnelling' through the barrier. This probability can be calculated for any given nucleus and the calculations agree very nicely with the experimental results. So α-decay is an effect of the combined strong and electromagnetic forces. γ-decay, the emission of an energetic photon from a radioactive nucleus, is an effect of the combined strong and electromagnetic forces. β-decay, the emission of an electron (sometimes a positron) and a neutrino from a radioactive nucleus is different. As it could not be explained by either of these forces (or by the gravitational force), a fourth force had to be involved – the weak nuclear force. This was postulated in 1934, by Enrico Fermi, a year before Yukawa's postulation of the strong force.

The typical β-decay process is the decay of the neutron into a proton, an electron and a neutrino, $n \rightarrow p^+ + e^- + v$. When the decay was first observed, physicists could 'see' the proton and

Fig. 2.5. A diagram of the potential energy versus distance from the centre of a nucleus. If an α-particle has an energy E inside the nucleus then, classically, it cannot get out. But quantum mechanics allows a 'tunnelling through the energy wall'.

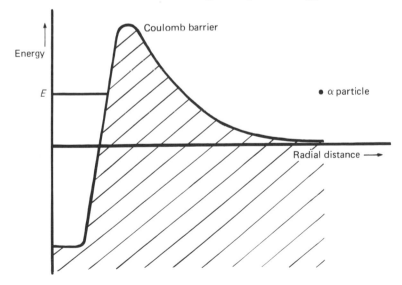

electron because both were charged and could leave tracks in a cloud chamber or effect a geiger counter. The third particle, the neutrino (*v*), was postulated by Wolfgang Pauli to make energy, momentum and angular momentum balance in the equation. At first the neutrino seemed an odd, almost unbelievable particle. It was supposed to have zero rest mass and zero electric charge. It carried energy and angular momentum, or spin. In the usual nuclear units, its spin is $\frac{1}{2}$. Physicists soon realised that this does not make it very different to the photon, which also has zero rest mass and charge and carries angular momentum (the spin of the photon is 1). Looking at this typical *β*-decay equation one can see that some new force must be involved, for neither the electron nor the neutrino is affected by the strong force and the neutrino is not affected by the electromagnetic force. And gravity, as usual is much too weak.

Another thing worth noticing about the decay of the neutron is that it takes a 'long' time – its half life is ~ 12 *minutes*. Now decays via the strong interaction typically take $\sim 10^{-23}$ *seconds*, so this new interaction is a weak interaction. It turns out also to be a very short range interaction – even shorter range than the strong interaction. So the particles that we suspect of 'carrying' it (as the pions carry the strong interaction, or photons the electromagnetic interaction) must be very heavy. They have not yet (1982) been found. They are called the intermediate vector bosons – and would benefit both from discovery and a shorter name. Also, because the range of the force is short the reaction time must be correspondingly short. We do not know just how short until we find this range, and the most obvious way of doing *that* is to find the mass of the intermediate vector boson; the heavier that is, the shorter the range of the force and the shorter the reaction time. It is probably less than 10^{-26} s.

This reaction time is not in any way the same as the *lifetime* of the decaying particle. Generally, we have three 'times': the production time of the particle, its lifetime, and the decay interaction time. The production time (if it is produced via the strong interaction) is around 10^{-23} s. The lifetime depends on the particle; it is $\sim 2 \times 10^{-8}$ s for a pion, around 10^{-23} s for many other sorts of meson that I shall deal with later in this chapter, ~ 12 minutes for a neutron, and either infinity or possibly some very long time like 10^{31} *years* for a proton.

The first theory of the weak interaction was developed by the famous Italian physicist, Enrico Fermi. He was a man equally good at theory or experiment. He won the Nobel Prize for physics in 1938 for discovering the production of new radioactive elements by neutrons. This led to his being in charge of the construction of the first nuclear reactor (in a squash court in the University of Chicago). I am a squash player and sometimes wonder, in times of acute political crisis, if the squash court might not have been put to better use by the squash club. I guess that very likely, in the future, mankind may be able to refrain easily and without effort from making unfavourable applications of knowledge. But I also feel that in the 1940s that was not at all possible and, if Fermi had not done it, someone else would. And it may well be that it was essential that he did. '*God works in a mysterious way His wonders to perform*'.

Fermi put forward his theory of β-decay (and the weak interaction) in 1934. He noticed that all the particles involved in the decay of the neutron were particles with spin (that is, intrinsic angular momentum) of $\frac{1}{2}$, in nuclear units. Such particles behave differently from particles like the photon (spin 1) or pion (spin 0). The particles with integral spin are called bosons (after Satyendra Bose, an Indian physicist). The particles with spin $\frac{1}{2}$ (or, in general, what is called half integral spin, $\frac{1}{2}$, $\frac{3}{2}$, $\frac{5}{2}$, etc.) are called fermions. One big difference between the two types of particle is that the fermions obey the famous Pauli Exclusion Principle. This states no two fermions can occupy the same quantum state. For instance, there is for any atom an electron orbit which is closest to the nucleus – the ground state, it is often called. Electrons have spin $\frac{1}{2}$. According to quantum mechanics, that spin can be 'up' ($+\frac{1}{2}$) or 'down' ($-\frac{1}{2}$). No other spin states are possible. We can put an 'up' electron and a 'down' electron in the lowest orbit, and despite being in the same orbit they are in different quantum states because of their spins. And that is that. No other electrons will go into that orbit. How do they know that they are forbidden to do this? That is the sort of question that has two answers. One is 'don't ask awkward questions', the other I shall deal with in chapter 6.

Fermi noted that in the decay of the neutron all four of the particles involved were fermions. He postulated this is *always* true in weak interactions. Now at once this seems to break down in the decay of the pion. Remember the pion decay is

$$\pi^{\pm} \rightarrow \mu^{\pm} + \nu^0.$$

Notice that the pion has spin zero so it is not even a fermion. But there is what looks like a sneaky way round this. (It is not really sneaky; it is true and very important.) This way says that the decay is not just $\pi \rightarrow \mu + \nu$ but

$$\pi^+ \rightarrow p^+ + \bar{n}^0 \rightarrow \mu^+ + \nu^0.$$

What is happening here is one of those 'virtual' transitions allowed by the Heisenberg Uncertainty Principle. The pion cannot decay to proton and an anti-neutron (that is what the bar over the n means) for any great length of time because the mass energy of the proton and anti-neutron is much greater (13 times) than the mass of the pion. But it can do so for a very short time Δt if $\Delta t \Delta E \approx h$, where h is Planck's constant and ΔE is the energy difference between the sum of the proton and anti-neutron's mass energies and the mass energy of the pion. And because it is allowed to do this, it does it. There is an often unstated principle in physics called the Totalitarian Principle. It says 'everything that is not forbidden is compulsory'. This is an example of its applicability. So a pion spends a certain fraction of its life as a proton, anti-neutron pair. Since Δt is *very* short compared to the pion lifetime it does this billions of times and is reconstituted safely as a pion. But eventually the weak interaction grabs it and the interaction $p^+ + \bar{n}^0 \rightarrow \mu^+ + \nu$ takes place. And that is the end of the pion. This Uncertainty Principle has all sorts of interesting applications. And, of course, each successful application of it (using this theory, for instance, to predict the lifetime of the pion) reinforces our belief that this is how the material universe really is. The pion is not always a pion, nor a proton just a bare stable, unalterable proton. All is change.

There is another odd thing about this weak interaction; it distinguishes between right-handedness and left-handedness. This was a very great shock to physicists when it was discovered in 1957. Even Wolfgang Pauli, who had worked with the weak interaction for half a lifetime and was a brilliant, witty man, was surprised. I have heard that he at once said 'this not only proves that God exists; it proves that He is a weak left-hander'. It is still very surprising that no small and apparently unstructured an entity as an electron can tell clockwise from anti-clockwise. I am not even sure that dogs can do that.

To explain this peculiar ability of the weak interaction I need to use the idea of reflection a lot, so it is as well to be clear what mirrors do. For instance, it is generally believed that mirrors reverse things left to right. If you write your name on a piece of paper and hold it up to a mirror then the reflected writing looks 'backward' – mirror writing. But really, mirrors do not do this. Two simple experiments demonstrate this. The first is to stand in front of a square mirror. Obviously, it does not reverse top and bottom – but, maybe, left and right. If it does this, just turn it through 90 ° and it should now reverse top and bottom, but not left and right. It does not. The other experiment is to write your name on a piece of transparent plastic and pick it up carefully, without reversing it in any way. Hold it up so that you can see the mirror through it and you will find that you can read your name on the plastic – and equally well on the reflection of the plastic. It is not the mirror that does the left to right reversal, it is ourselves. (Any psychologists might like to consider why almost everyone projects this apparently innocuous habit of reversing onto the mirror – 'it's those horrible mirrors that do it – reversing things left to right, confusing people – nasty perverted things, they are. And they are all alike – should be smashed, every one, or, at least, covered up with black cloth'. To me, the simplicity of the act hardly seems to justify the near universality of the projection.)

What mirrors really reverse is front to back. It is easy to see this if one takes an arrow and holds it in front of a mirror. If you hold it with its head up, then the reflected image has its head up; if the head is to the right, the reflected image has its head to the right. But if you hold it with the head pointing away from you and towards the mirror, then the reflected image has its head pointing *towards* you.

Before 1957, physicists thought the laws of physics would be the same in the looking glass world as in our real world. They had a lot of good reasons for believing thus. If one looked at experiments in the mirror they gave the same results as in the real world. Things still fell downward – they had the same acceleration due to gravity, and their positions and velocities obeyed the same formula. So too with optical experiments. A looking glass mirror still has its angle of incidence equal to its angle of refraction. A looking glass prism still breaks white light up into a rainbow spectrum. Same again with heat experiments and electrical experiments: looking glass magnets attract looking glass iron filings and so on. So physicists concluded that the laws of nature are the same in the mirror world. Technically, we assumed the laws were symmetric under reflection of the spatial co-ordinates (that is a bit stronger than is the case with most mirrors which, as we have seen, reverse only one co-ordinate, front to back). We did not stop to consider that *all* the effects and laws we had observed dealt with only two of the force fields – electromagnetism and gravity. Can we see if this also holds for the weak interaction?

It turns out that we can. The weak interaction shows up particularly in β-radioactivity. So we consider a very simple sort of β-decay, the decay of the muon, $\mu^+ \rightarrow e^+ + v + v$; it decays to an electron and two neutrinos. Suppose we have a muon moving along the line A to B in fig. 2.6. Suppose also that the spin axis of the muon also lies along the line with the rotation having the sense shown. If we look at the muon from point A the rotation appears to be anti-clockwise; if we look at it from B it is clockwise. Now suppose the muon decays. The electron that is produced may fly off in any direction. If we looked at a large number of such decays then, before 1957, we would have expected as many electrons to be in the forward hemisphere (that is, the hemisphere from which the muon appeared to be rotating in a clockwise direction) as in the backward direction. If that had happened then the result in the case of the mirror muon would have been the same and, once again, the laws of physics would have turned out to be identical in both worlds. *But*, in our real world it is not like that; the electron shows a distinct preference to go off into the forward hemisphere, which is the hemisphere from which we saw a clockwise rotation of the muon. And so in the mirror world, the mirror electron prefers to go into the hemisphere from which the mirror muon's rotation had been anti-clockwise. The laws of physics are different – or, as we physicists say, parity is not conserved.

Hints that this might be so turned up in the early 1950s when people were puzzling over the new charged mesons (what we now call kaons). At first, it was possible to believe that each new decay scheme meant a new type of particle. Two of these were called the τ-meson and the θ-meson. The τ-meson we have already come across; its decay scheme is $\tau^+ \to \pi^+ + \pi^+ + \pi^-$. The θ-meson had a simpler scheme, $\theta^+ \to \pi^+ + \pi^0$. The two right-hand sides to these schemes have different parity (which is

Fig. 2.6. This diagram shows a muon moving from point A towards a mirror (with its reflected image moving in the opposite direction). The muon is spinning; if viewed from A the spin is anticlockwise. If the muon decays the electron can fly off in any direction; one particular direction and its reflected image are shown.

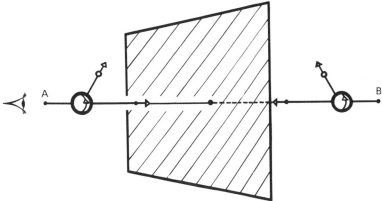

positive or negative). Before 1957, this was good evidence that we were dealing with two different particles. But as techniques improved and accelerators came into the picture the experimentally-determined masses and lifetimes of the two got closer and closer. The masses became 966.7 m_e for the θ and 966.3 m_e for the τ; the lifetimes 1.21×10^{-8} s for the θ and 1.19×10^8 s for the τ. At this point two Chinese-American physicists, Tsung Doo Lee and Chen Ning Yang threw in the towel. They had a further look at the evidence and found that, as far as weak interactions were concerned, there was *no* evidence that parity was conserved. They suggested some other experiments to test this possibility. One was carried out by another Chinese-American, Chieng-Shiung Wu and her co-workers. They placed a sample of radioactive cobalt in a strong magnetic field and very low temperatures. At low temperatures the atoms were almost at rest and their nuclei could be lined up (each nucleus is a small magnet) by the magnetic field. We then have a system very similar to our muons. Wu and her collaborators found that more of the β-decay electrons came off in one direction than in the other. Yang and Lee shared the Nobel Prize for physics in 1957.

So the weak interaction distinguishes between left- and right-handedness. We find that we are in a left-handed universe (hence Pauli's remark). Neutrinos are left-handed and anti-neutrinos are right-handed. One can tell the handedness of a neutrino because it has spin and a direction of motion so one can envisage it as a screw – its thread is left-handed.

There are three simple operations that we can carry out on a system of particles. The hardest to understand is the one we have just dealt with, the parity transformation, called P for short. This operation reverses the sign of all co-ordinates, length, breadth and height, roughly. If we have a room and take one bottom corner as the origin (that is, the place from which we make our measurements) then we can specify the position of any object in the room by giving its height above the floor and its perpendicular distance from the two vertical walls that meet the floor and the origin – three numbers x, y and z, say. If we apply the parity transformation to these they become $-x$, $-y$, $-z$. I have done this in fig. 2.7. Again, you can see why physicists were fooled by the weak reaction for so long. It is not obvious that this simple operation should alter the laws of physics.

The second of the simple operations is called charge conjugation, C – just changing the sign of all charged particles in the system: positives to negatives, negatives to positives. The third is time reversal, T – just reversing the direction of time, the sequence in which events happen. (I saw a definition of time once in *Reader's Digest*, or some such magazine, 'time is just nature's way of making sure that everything does not happen at once'. I forget to whom it was attributed – maybe Woody Allen. I found it very funny because it seemed to be taking the mickey out of a portentous sort of cliché. Now I begin to wonder if it may be true.) Figure 2.8 shows an original interaction, a negative pion hitting a stationary proton to produce a Λ^0 hyperon and a K^0 meson, and the result of applying P, C and T to it.

Now, let us go back to the original muon decay (fig. 2.6), which showed that parity is not conserved in the weak interaction. That is, when we reflected the system, the laws of nature were different. Suppose we not only reflect the co-ordinates but also change the sign of the charges. Then we have not only a mirror image but a mirror image in which all the positive charges have

Fig. 2.7. The parity transformation.

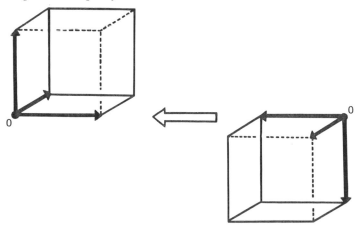

become negative and vice versa. When we examine this in detail we find that the laws of nature are *not* altered (in the case of the muon decay and, indeed, in the case of most weak decays). So although the system is altered by the application of P, it is not altered by CP. (Physicists say 'it is invariant under the application of CP'.) This is very interesting; when we are dealing with the weak interaction, changing the co-ordinates changes the laws of nature. But changing both the co-ordinates and the signs of the electric charges leaves them unaltered. Amazing.

A very general theorem suggests if we apply all *three* of the transformations, that is, PCT, then the laws should stay the same; they are invariant under PCT, to use physicists' jargon. So if PC leaves them unchanged, T must leave them unchanged. *If we reverse the direction of time nothing changes.* This applies to particle physics where we are dealing with individual particles. With large collections of objects seemingly another effect sets in. But for these small particles, and few of them in any one interaction, it seems that the direction of time is unimportant. Except in one small case. To follow that we need to look at the neutral kaons, and to understand their weird behaviour we must introduce a new concept. The behaviour of those particles seemed so odd to the physicists who investigated them that they

Fig. 2.8. The three operations P, C and T applied to the reaction $\pi^- + P^+ \rightarrow \Lambda^0 + \bar{K}^0$. The original reaction is shown on the left and the effects of applying each of the operations to it are shown on the right.

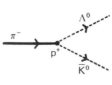
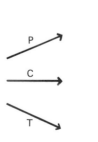
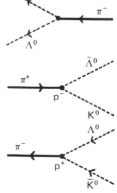

called this property 'strangeness' and gave the particles the generic name 'strange'.

I feel that it is appropriate at this point to summarise what we know about the four force fields of the physical world. This I have done in the accompanying table. It is worth noting some points. One is the exceptional weakness of the gravitational force. Yet it is the force that holds the Earth in its orbit, the Galaxy and the clusters of Galaxies together. In black holes it overcomes all other known forces. Other things to note are the ranges and that two of the carriers have not yet (September 1982) been discovered.

We can now look at how these strange particles earned their odd name – this business of associated production. You remember it was postulated that in strong, productive reactions, particles like the Λ^0 hyperon and K mesons are always produced in pairs. Hence, when they fly off outside the strong force range of each other they cannot decay via the strong force (in about 10^{-23} s) but must decay via the weak force (in 10^{-10} to 10^{-8} s). Pais and others explained this by postulating a new property called 'strangeness'. Now this sort of thing had been necessary earlier on in physics. When people first became acquainted with electricity they had to postulate a property called 'charge' (why charge?), also to assume that objects could have either a 'positive charge' or a 'negative charge'. They found, too, that if two bodies both had a positive charge they repelled each other, but if the charges were different they attracted each other. Fermi took over the idea in his theory of the weak force and postulated 'weak charges'. Pais now postulated 'strangeness' and said that the particles we knew (at that time), the proton, neutron and pions, had zero strangeness. But the kaons and Λ^0 had strangeness ± 1, depending on whether they were particle or anti-particle (just as the electron has electric charge -1, and the positron, $+1$). Other particles have even greater strangeness.

We can then explain the 'associated production' of 'strange' particles by saying that strangeness must be conserved in strong interactions (just as is parity). So a strong reaction like $\pi^- + p^+ = \Lambda^0$ is forbidden because the strangeness on the left hand side is $0+0=0$, but the strangeness on the right hand side is -1. However, $\pi^- + p^+ = \Lambda^0 + k^0$ is allowed because then the strangeness on the right hand side is $-1+1=0$.

In the *weak* reaction, strangeness (again like parity) is *not* conserved, so that $\Lambda^0 \to p^+ + \pi^-$ is allowed. This means we have discovered a new property of strongly interacting particles (hadrons is their generic name) and determined its value for all the different particles. This property is conserved in strong interactions but not in weak interactions. In this it is like parity and unlike electric charge, which is always conserved (well, as of September 1982). Having gained knowledge of this property we could examine various situations, assess them for strangeness, and predict (in some cases) previously unexpected behaviour. The neutral kaons are a particularly interesting case.

Consider again what is almost our 'standard' interaction, $\pi^- + p^+ \to \Lambda^0 + k^0$. Λ^0 has strangeness of -1, π^- and p^+ have zero strangeness. It is a strong reaction, so strangeness is conserved and so k^0 must have a strangeness of $+1$. Now we can also have a reaction $\pi^+ + p^- \to \overline{\Lambda^0} + \overline{k^0}$, where the bars over a letter mean an anti-particle. Obviously the neutral anti-kaon

Table 2.1 *The physical force fields.*

Force	Relative strength	Range in m	Carrier	Nature
Strong	100	$\sim 10^{-15}$	pions and other mesons	attractive at longer distances; repulsive core
Electro-magnetic	1	infinite	photons	attractive for unlike charges; repulsive for like
Weak	10^{-12}	$\lesssim 10^{-17}$	intermediate vector bosons?	attractive for unlike weak charges; repulsive for like
Gravity	10^{-37}	infinite	gravitons?	attractive

has strangeness of -1. This is the *only* difference between the neutral kaon and its anti-particle. For pions, because they have zero strangeness, the neutral anti-pion is *identical* with the neutral pion; the π^0 is its own anti-particle. However, k^0 and $\bar{k}^0$ have this difference of strangeness so they are distinct.

What can we say about this? Well, strangeness is *not* conserved in decays, so both k^0 and $\bar{k}^0$ can decay into a π^+ and π^-, $k^0 \to \pi^+ + \pi^-$ and $\bar{k}^0 \to \pi^+ + \pi^-$. But this means that we can have transitions from k^0 to $\bar{k}^0$, via a weak virtual interaction, $k^0 \to \pi^+ + \pi^-$ (virtual) $\to \bar{k}^0$. So neither the k^0 nor the $\bar{k}^0$ can exist in a pure state – they are always flipping from one to the other. It is a rather bizarre quantum-mechanical effect and needs to be treated by quantum mechanics. When one does this one finds that one cannot predict a definite lifetime for either the k^0 or the $\bar{k}$. But states exist, which have been called k_1^0 and k_2^0 that are mixtures of the k^0 and $\bar{k}^0$ states in a definite, well-known way. These two have definite mean lifetimes and they are quite different! The k_1^0 has a mean life of 0.83×10^{-10} s. The k_1^0 decays to two pions. This was the mode seen by Rochester and Butler in their original picture (fig. 2.1). The k_2^0 decays to *three* pions, typically, $k_2^0 \to \pi^+ + \pi^- + \pi^0$, and cannot decay to two pions without violating CP invariance. 'CP invariance' means that the laws of physics remain unchanged if we reflect all the co-ordinates *and* change all positive charges to negative and vice versa.

This very odd state of affairs was *predicted* once strangeness was discovered. It was fairly easy to check. By that time, 1956, accelerators were sufficiently energetic to produce kaons. Many examples of the short-lived neutral kaon had been seen. If the theory was right, a detector could be put about 600 times further away and it would pick up the long-lived neutral kaons. In practice, the detector needs to be far enough away from the target to make sure that no short-lived k_1^0 can reach it. A team from Columbia and Stonybrook did this at the Brookhaven Cosmotron. They used a Wilson cloud chamber as detector and, sure enough, found the long-lived k_2^0 and showed that it did indeed decay to *three* pions. Quantum mechanics may seem at first sight to be an odd sort of theory, but it works.

However, things were even odder than this. I have stated that the long-lived k_2^0 decays to three pions and that it cannot decay to two pions only without violating the double symmetry CP. It turns out that about two decays in every thousand do just this. There are now very solid reasons for believing that if we apply all three operators, CPT, thereby reversing the spatial co-ordinates, the electric charges and direction of time, the laws of physics are unaltered. If that is true and if CP is violated, then T, time reversal symmetry, would be violated in a balancing way. So these k_2^0 mesons, small and apparently uncomplicated as they are, 'know' which way time is going. This does not seem to be a macroscopic effect like the entropy increase in the world at large. It was a very surprising discovery and Cronin and Fitch were awarded the Nobel Prize for it in 1980.

This again was a discovery made in an accelerator laboratory. Since about 1954, most of the discoveries in particle physics have been made in accelerators. So I will backtrack again and say something about accelerators.

Nowadays most people have an accelerator in their homes – it is the tube of their TV set. If it is a colour set it takes electrons from a heated wire and accelerates them by applying a static electrical potential of about 10000 volts between the 'electron gun' and a screen with a hole in, called the anode. The electrons go through the hole and hit the screen of the TV tube, which is coated with phosphors that glow with various colours when hit.

This straightforward application of an electrical potential was the way the earliest accelerators for nuclear bombardment were made; the Cockcroft–Walton generator and the Van de Graaff machine. Initially, these worked between 100000 electron volts (100 keV) and a million electron volts (1 MeV). Nowadays they can reach energies ten or more times higher. These energies are useful in *nuclear* physics, the study of the way nuclei are constructed. But they are too low for particle physics. For this the projectile needs enough energy to create at least one pion $[m_\pi = 140 \text{ MeV}]$.

The first real breakthrough in accelerator design was made by Ernest Lawrence at Berkeley, California. He was awarded the Nobel Prize for it in 1939. His idea was to apply a number of 'small' accelerations to the projectile, instead of the one big drop used by Cockcroft, Walton and Van de Graaff. He was able to do this by using a strong magnetic field to bend the trajectory of the particle into a roughly circular path. This path was arranged to be between two D-shaped electrodes, as in fig. 2.9. An alternating voltage of some quite 'low' ($\sim 30\,000$ volts)

Fig. 2.9. A diagram of a cyclotron.

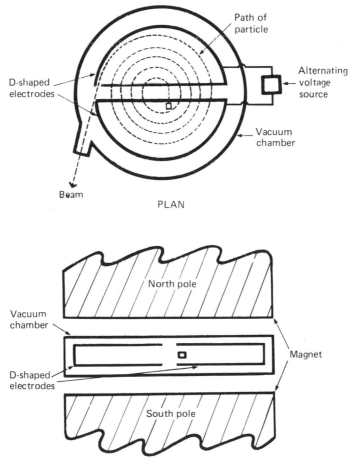

value was applied to the D's and the frequency chosen so that each time the particle got to the gap it received an acceleration. It is the same principle as pushing a child on a swing – a series of little pushes given at just the right moment works just as well as one very big initial push. Since the particle is constantly gaining energy its path is not a circle but a spiral. Eventually, when it is energetic enough, it reaches the outside edge and then can be persuaded to move out of the magnetic field and strike whatever target we choose. Lawrence called this device a cyclotron. The cyclotron worked because the time a particle takes to make a half circle is independent of the radius of the circle. Slow particles have only a small half circle to travel; as they speed up, the half circle gets bigger. So all the particles arrive at the gaps between the D's at the right time to get their impulses.

At least, that is what happens at low energies. As the energies get higher the projectile, according to the theory of relativity, gets more massive; the observed mass m is related to the rest mass m_0 by a simple formula $m = \gamma\, m_0$. γ is called the Lorentz factor after one of the early pioneers of relativity theory. It is given by $\gamma = 1/(1-v^2/c^2)^{\frac{1}{2}}$ where v is the velocity of the particle and c is the velocity of light. One can see from this that as the particle velocity v gets close to c, v^2/c^2 gets close to unity, so that $(1-v^2/c^2)^{\frac{1}{2}}$ becomes small and hence γ, its inverse, becomes large; m gets bigger than m_0 and keeps on getting bigger as v approaches c. The particle gets heavier and heavier and begins to lag behind. It begins to get out of step.

It was quite a surprise to find what had been thought to be rather esoteric, relativistic effects creeping into engineering. But the engineers adjusted admirably (with help from physicists). It was found that one could compensate for this increase in mass by adjusting the magnetic field and the frequency of the accelerating voltage to suit the particles as they accelerated. One could no longer have many bunches of particles in the machine, at different radii at the same time – one bunch had to make the complete trip before the next could start. That was a minor disadvantage (and could even be used with profit, sometimes).

As the machines became bigger and people became expert at varying the magnetic field in suitable ways, the magnet became a ring magnet rather than a solid cylinder of steel. These ring magnets are now several *kilometres* in circumference in the largest accelerators. Figure 2.10 shows a photograph of part of the ring magnet at Fermilab, near Chicago. This accelerator can accelerate protons to $500\,000\,000\,000$ eV (5×10^{11} eV or $500\,\text{GeV}$)! (May 1982). The Fermilab engineers are hoping to *double* this energy in the near future. At 500 GeV a proton is just over 500 times flatter, in its direction of motion than a slow proton and more like a very flat, particularly dense pancake than a sphere. And if it were radioactive its lifetime in our frame of reference (frequently called the laboratory frame) would be 500 times longer. That is another effect the engineers have to allow for when they build these large accelerators and make provision for delivery of beams of pions, kaons, etc., to various experimental halls.

It was at a Chicago accelerator that the first of the so-called 'resonances' was observed. Once again, Enrico Fermi was involved. The Chicago group determined the 'cross section' for the interaction of a positive pion with a proton at various energies of the pion (the protons were at rest in the laboratory). Imagine firing the pion at an area of one square centimetre containing one proton. Most of the square centimetre is, of course, empty. But one small area is effectively filled by the proton and *that* area is the cross-section available for interaction. To determine it experimentally of course, one needs many more protons. Also, it is different to the billiard ball case – the area varies with the pion energy. Fermi and his co-workers found that, in the pion energy range they were using, it varied remarkably. The graph is shown in fig. 2.11. To a physicist, this immediately suggests that some system is resonating.

Resonance is another phenomenon typically associated with wave motion. It is particularly easy to demonstrate with sound. If a guitar string is tuned to a particular frequency and the same note is played on a violin near by, the guitar will also sound. If one approaches the note on the violin gradually, the guitar

Fig. 2.10. Part of the giant 500 GeV accelerator at Fermilab. The ring magnet is in a tunnel under the low ring mound, which is itself between the canal and the road. The canal and fountains are used for cooling the water which cools the magnets (and for canoe racing in summer).

will not sound to begin with but, as the two frequencies get closer, it will begin to 'resonate'. Caruso was famous for doing this with a wine glass. His pitch was so true and his volume so great that he could make a wine glass resonate so strongly it shattered. In the case of the guitar string we could measure the amplitude of the vibration induced in the string for various values of the frequency of the violin note. We would then get a curve like the curve of fig. 2.11, with a peak at the frequency to which the guitar is tuned. The width of the curve gives us a visual idea of the sharpness of the resonance. This sharpness depends on the amount of frictional damping of various sorts to which the guitar string is subjected.

In the case of the pion–proton interaction we are encountering a quantum-mechanical wave phenomenon. What corresponds to the tuned guitar string is a new particle now called the delta (Δ^{++}) hyperon. Its mass, 1236 MeV (the proton mass is 976 MeV), is the equivalent of the guitar string frequency. When the combined energy/mass of the pion–proton system

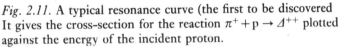

Fig. 2.11. A typical resonance curve (the first to be discovered It gives the cross-section for the reaction $\pi^+ + p \rightarrow \Delta^{++}$ plotted against the energy of the incident proton.

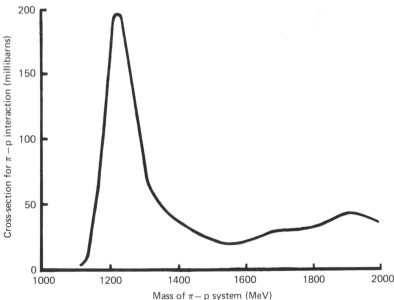

reaches this value, the system 'resonates' and there is a very high chance of this particle being produced:

$$\pi^+ + p^+ \rightarrow \Delta^{++}.$$

This new particle has an electric charge of $+2$ units; the first of the strong particles we have come across that has a charge greater than unity.

But it has another property that distinguishes it from the hyperons we have come across so far: its strangeness is zero. So it does not need a balancing particle to be produced at the same time *and* it can decay via the strong interaction. It is produced in $\sim 10^{-23}$ s, lives for about the same time – and then decays. It is very ephemeral, but does live long enough for us to measure its various parameters. We now know its mass, spin ($\frac{3}{2}$), charge ($+2$), even its very short lifetime. The latter we get from the Uncertainty Principle. We can measure the energy width, E, of the resonance peak and use $\Delta E \, \Delta t \geqslant h$ to get this lifetime. You can be certain it is right; after all, we used the Uncertainty Principle to measure it.

Soon many more of these short-lived particles were discovered (and continue to be discovered). The only difference between them and the longer-lived particles is, because of their properties, they can decay via the strong interaction. They come in two main classes, baryons and mesons. The proton is a typical baryon and the pion, a typical meson. There is yet another property, called baryon number, which distinguishes these two categories. Baryons are particles with a baryon number of $+1$; mesons have a baryon number of zero; anti-baryons, like the anti-proton, have a baryon number of -1.

But even with this simplification we still face this enormous 'zoo' of what we had once thought to be 'fundamental' particles. 'Zoo' is a good word. In a zoo one finds all sorts of animals, some very large, such as elephants, and some small, such as humming birds. Tortoises are long-lived, butterflies not. Some are nocturnal and hard to see; some are just exhibitionists. Others are so strange that they are hardly credible, such as giraffes or duck-billed platypuses. But, returning to the particles, it is hard to believe, given there are so many of them, that they are 'elementary', or 'fundamental' in the sense of Daltonian atoms. In fact, the situation is much the same as with the chemical 'elements'. The ancient Greeks needed only four (or maybe five) of these. Dalton needed (eventually) 108 or more. By 1962 the situation with the 'fundamental' particles was similar; there were well over a hundred of them. Could it be that these 'fundamental' particles were not fundamental at all, but rather made out of something simpler?

3

Quarks

Table 3.1 gives the names and some of the properties of a few of the particles whose existence had been established by 1962.

Table 3.1 *A list of some of the particles, and some of their properties, known by 1962. Anti-particles in general are not listed.*

Particle	Rest mass in MeV	Charge	Lifetime in seconds	Spin in units of $h/2\pi$
Photon	0	0	∞	1
Electron	0.51	-1	∞	$\frac{1}{2}$
Neutrinos	0	0	∞	$\frac{1}{2}$
Negative muon	105.6	-1	2.2×10^{-6}	$\frac{1}{2}$
Negative pion	139.6	-1	2.6×10^{-8}	0
Neutral pion	135.0	0	0.9×10^{-16}	0
Negative kaon	493.8	-1	1.2×10^{-8}	0
Neutral kaon	497.8	0	$\begin{cases} 0.9 \times 10^{-10} \cdot \\ 5.4 \times 10^{-8} \end{cases}$	0
η meson	548.8	0	2.5×10^{-19}	0
π meson	765	0	5.1×10^{-24}	1

plus many heavier, short-lived mesons.

Particle	Rest mass in MeV	Charge	Lifetime in seconds	Spin in units of $h/2\pi$
Proton	938.3	$+1$	> 3.10	$\frac{1}{2}$
Neutron	939.6	0	932	$\frac{1}{2}$
Λ^0 hyperon	1115.6	0	2.5×10^{-10}	$\frac{1}{2}$
Σ^+ hyperon	1189.4	$+1$	0.8×10^{-10}	$\frac{1}{2}$
Σ^0 hyperon	1192.5	0	1.0×10^{-14}	$\frac{1}{2}$
Σ^- hyperon	1197.3	-1	1.6×10^{-10}	$\frac{1}{2}$
Ξ^0 hyperon	1314.7	0	3.0×10^{-10}	$\frac{1}{2}$
Ξ^- hyperon	1321.3	-1	1.6×10^{-10}	$\frac{1}{2}$
Ω^- hyperon	1672.4	-1	1.3×10^{-10}	$\frac{3}{2}$
Δ^{++} baryon	1236	$+1$	5×10^{-24}	$\frac{3}{2}$

plus many other short-lived baryons.

There are one or two points to make about this table. I have listed 'neutrinos' because there are several sorts of neutrino, one 'belonging' to the electron, another to the muon and so on. They are distinct only in properties not listed in the table. Also, I have listed the neutrino rest mass as zero, which I think is the likely value. But this has not yet been established for certain (May 1982). A few particles that probably exist, for example the graviton and the intermediate vector bosons, I have not listed. And many short-lived mesons and baryons are not listed by name. In all, well over 100 of these are known.

Thus in 1962 we had come a very long way from the apparently elegant simplicity of 1947. With so many particles, and none obviously more fundamental than any other, it was hard to believe that these are the fundamental building blocks of nature. But some simplification and systemization is possible. For instance, the particles can be classified into a few broad categories. There are the leptons (literally, the light ones – particles of small mass). The known leptons in 1962 were the electron, the muon and their respective neutrinos (written v_e and v_μ). It had been a bit of a shock to discover that the neutrino emitted in the decay of a pion was different to the β-decay neutrino. At the time it seemed an unnecessary complication (but then, so did the existence of the muon). Nowadays we know that the situation is even more complicated. There is at least one lepton, the tau (τ) particle, heavier than the muon and *it* has its own neutrino too (the v_τ).

Apart from the leptons there is the photon, maybe the graviton, and the really prolific class, the hadrons (meaning the strong ones); particles that interact via the strong interaction, as well as the other interactions. Hadrons themselves come in two types – mesons (with integral spin) and baryons (with half integral spin). The pion is a typical meson; the proton is a typical baryon.

We describe these particles by properties, some of which are quite familiar from ordinary life, others not so. I have discussed some of these already, for example strangeness. Still others I shall mention soon. In the first category of properties are mass, lifetime, electric charge (positive, negative or neutral), and intrinsic angular momentum (called spin). Mass and lifetime are 'secondary' properties. Either we derive them from the theory of particles we have, or expect to be able to derive them from a complete theory. Spin is different; it is a very fundamental property. It is measured in units of $h/2\pi$ (often written $\hbar$) where h is Planck's constant. Particles whose spin is an integral multiple of $\hbar$ (that is, whose spin is 0, $\hbar$, $2\hbar$, etc.) are called bosons. The mesons are bosons (as you can seen from table 3.1). Bosons do *not* obey the Pauli Exclusion Principle. The other type of particle, the fermions, have half integral spin (that is, their spin is $\frac{1}{2}\hbar$, $\frac{3}{2}\hbar$, $\frac{5}{2}\hbar$, etc.) and obey the Pauli Exclusion Principle (that is, small as they are, they can read a 'house full' sign and obey it). The baryons and leptons are fermions.

Some people find it difficult to grasp that spin is both intrinsic and extremely important. Our ideas of angular momentum derive from our experience in the macroscopic world. We know that a whirling top has spin (when it is going). We can see that this spin alters its properties. When it has enough spin it will stand upright on its point. When it is really spinning it becomes very stable, and in addition to its fast rotation it may have a much slower gentle precession.

The Earth itself spins about its axis and has a great deal of angular momentum. This spin is not so obvious as that of the top; in fact for many millennia it went unnoticed by most people. But nowadays we believe in it and are able to use that belief to calculate consequences – predict sunrises, eclipses, satellite orbits, and so on. But neither the Earth's spin nor the top's spin appears to be quantised. It looks to us as if we could make the top spin just as slowly as we wanted. But this is not so – angular momentum is a quantised quantity but, as with energy, it is quantised on a very small scale. $\frac{1}{2}\hbar$ is a very small amount of spin.

If we make a measurement of spin the smallest answer we can get is $\frac{1}{2}\hbar$ (for fermions) or, for a boson with non-zero spin, $\hbar$. This sounded crazy to classical physicists when they first came across it, but that is how it is in the quantum world (and that is the physical world).

Then there are the properties I have mentioned before which have no analogue in classical physics. Baryon number and lepton number are two of these. All baryons have non-zero baryon number (1, 2, etc.) and all non-baryons have zero baryon number. The same goes for lepton number, except that nowadays we know that it comes in these various 'flavours' – electron lepton number, muon lepton number and so on. Strangeness is another quantum number I have already mentioned. It distinguishes the 'strange' hadrons from the 'ordinary' hadrons (like the proton and pion) and even classifies the strange particles into 'strange', 'very strange' and 'exceptionally strange'. (As I write this down it all seems so strange as to be hardly credible; no wonder Heisenberg got upset by early quantum mechanics.)

Then there are a number of properties I have not so far mentioned. One of these is called isotopic spin, or isospin for short. It was first noticed shortly after the neutron was discovered in 1932. Once the neutron was known it soon became obvious that, apart from their difference in electric charges, it and the proton were very similar particles. They seem to behave identically as far as the strong force goes. For instance, hydrogen has three isotopes. Ordinary hydrogen has a nucleus consisting of a single proton; heavy hydrogen, sometimes called deuterium, has a nucleus made of a proton plus a neutron. The third isotope, which is radioactive and called tritium, has one proton plus two neutrons. Normal helium has a nucleus consisting of two neutrons plus two protons. But it has an isotope, helium 3, consisting of one neutron and two protons. We can measure the binding energy of the nuclei of both helium 3 and tritium. We find that, once we have made allowance for the electrostatic repulsive force of the two protons in helium 3, the binding energies are identical. So the strong force between two protons and a neutron is identical with the strong force between two neutrons and a proton. As far as the *strong* force is concerned neutron and proton are the same particle. Often, when we are dealing with only the strong force, we just call them both nucleons. The strong force is 'charge blind'. But, of course, the electromagnetic force is not, and there is a charge difference between the proton and the neutron. One has charge $+1$, the other zero. Also there is a small mass difference of 1.3 MeV. (Notice that in table 3.1 I have taken Einstein and $E = mc^2$ seriously, and given the masses in units of energy, MeV.)

Now a proton has spin $\frac{1}{2}\hbar$. It is also charged. The combination of charge and spin means that it is, in effect, a small magnet. (It was a bit of a surprise to find the *neutron* also has a magnetic moment. This straight away tells us that the neutron is not a simple, 'elementary' particle. It must have a charge distribution in and around it, even though its total charge is zero.) Getting back to the proton, this means if I place a proton in a vertical magnetic field it can sit in one of two states, with its magnetic axis either parallel to the magnetic field, or anti-parallel. Crudely, with its axis *up* or *down*. Since the spin is quantised these are the only two possible positions. Now in these two states its energy is very slightly different. When Condon and Cassen were considering the proton/neutron similarities and differences they saw the similarities to this case and suggested there was another quantum quantity, isospin, that separated the two states of the nucleon (that is, the proton and the neutron) just as the *spin* separated the two states of the proton in the magnetic field. They found this works – the isospin behaves mathematically just like spin. But it does *not* have the mechanical effects we are used to in 'ordinary' spin. Isospin resides in its own space, not our ordinary 3-dimensional space. Also, it is fairly obviously connected to electric charge – because that is one of the obvious differences of a proton and a neutron.

So in 1962 we had these many particles, separable into broad categories, and the category with by far the most particles was the hadrons. The hadrons themselves are divisible into two main groups, the mesons and the baryons, and we describe them using the properties of mass, charge, spin, lifetime, baryon number, lepton number(s), strangeness, and isospin. Later on, we shall need more properties but, for the moment, these will do.

Physicists next looked for symmetries amongst these properties. They noticed that symmetries and conservation laws are connected. In fact, some of our most basic conservation laws can be derived from just considering the symmetries implied in the idea of 'empty space'. So before I get on to the symmetries in particle properties I shall have a look at symmetries in general.

Suppose we have three simple geometric shapes, a square, an equilateral triangle and a circle, on a sheet of paper. If we move any one of them to one side or up and down, it remains a square, or whatever. That is to say, it is invariant under lateral or vertical displacement. Or, in short, pushing it around does not alter it. Of course, in saying this I have made some assumptions about our square. I have assumed it is made out of stiff cardboard, or sheet steel – something pretty rigid. If I had made my square out of thin rods of plasticine it would *not* have been 'invariant under lateral displacement'. But let us stick with rigid shapes.

Suppose now we left it for five minutes. Again, it would not change. So it is invariant under temporal displacement.

Now suppose we notice that the sheet of paper has an edge – or better still, a preferred direction: up, or north, or whatever. Suppose we now rotate the square. If we rotate it through 20 ° it looks different; its edge is at a different angle to the preferred direction. But if we rotate it through 90 ° it has not noticeably altered. So the square is invariant under rotations of 90 ° (or multiples of 90 °). The triangle is invariant under rotations of 120 °. And the circle is invariant under *any* rotation. So the circle is more symmetric than the square or triangle. And notice that the circle has a sort of continuous symmetry, whereas the symmetries of the square and triangle are discontinuous.

Now consider empty space. If we move in any direction, nothing changes. It is invariant under displacement. It is the same in all directions. If we imagine some particle placed at a point in empty space, it will stay there, for if it moved in any one direction that would be a preferred direction. Because there is no preferred direction, it stays put; its momentum does not change. So conservation of momentum arises from empty space being invariant under displacement.

Similarly, conservation of energy arises from translation symmetry in time, and conservation of angular momentum from the symmetry of empty space under rotation.

In 1960 people began to notice symmetries in the new particles. Murray Gell-Mann of Caltech and Yuval Ne'eman of the Weismann Institute, Israel, arranged the eight hyperons and nucleons as shown in fig. 3.1.

Spin is what is called a vector quantity – it has direction as well as magnitude. If a sphere is spinning it must be spinning about some axis and in one of two directions about that axis – left-handed or right-handed. So to describe spin completely one must state not only the magnitude of the spin but also these directions. One can represent it by an arrow – its length gives the magnitude and its direction. But any such arrow can also be represented by three components, as shown in fig. 3.2. So spin can be represented this way – and so can isospin. In particle physics there is a conventional way of looking at isospin and its three components, so that one need only state the complete value of the isospin and the magnitude of one

Fig. 3.1. The family of nucleons and hyperons displayed on a plot of strangeness versus isospin projection. The symmetry produced is obvious.

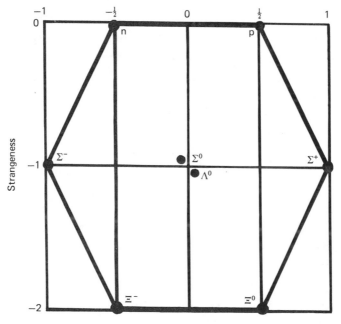

Isospin projection

component to describe it. The third component is the one conventionally chosen.

What Gell-Mann and Ne'eman chose to do was to plot this third component of the isospin against the strangeness for the eight particles, proton (p), neutron (b), and the Λ^0, Σ^+, Σ^-, Σ^0, Ξ^- and Ξ^0 hyperons as in fig. 3.3. The symmetry of the diagram is obvious and by inspection we can see a simple relationship between electric charge, Q, the third component of isotopic spin (I_3), the baryon number (A), and the strangeness (S). It is

$$Q = I_3 + A/2 + S/2.$$

All the particles in this diagram have a value of the 'ordinary' spin of $\frac{1}{2}$. It turned out to be even more insightful to move to a 'family' of 'resonances' with spin $\frac{3}{2}$. At that time (1962) all the known members of the family were short-lived particles. There is obviously a missing member in this family. It should have a I_3 value of 0 and a strangeness of -3. Now this is an interesting situation. All the other members of the family can decay via the strong interaction, but the new particle (which was christened the Ω^-) cannot. To begin with, its electric charge, from the formula, is -1. Also, from the diagram, it has this exceptionally high value of strangeness, -3. If we look at the

Fig. 3.2. The x, y and z components (x_1, x_2 and x_3) of a vector x.

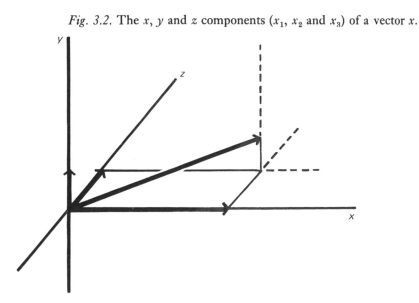

masses of the various rows in the diagram they increase downwards from the top row by increments of 144 MeV. So if an Ω^- decays into a Ξ^* (which gets rid of two units of strangeness) then it still has to find a particle with strangeness -1 and mass less than 144 MeV. No such particle exists. It is the same with all other configurations. Ω^- cannot decay and conserve strangeness, mass, etc. So it cannot decay via the strong force. But weak decays do not conserve strangeness. However, they also take a 'long' time. So this new particle should be, not a short-lived particle like the rest of its family, but a comparatively long-lived particle with a lifetime comparable with most of the hyperons in fig. 3.3, $\sim 10^{-10}$ s.

So, just from symmetry, Gell-Mann and Ne'eman were able to predict the properties of this as yet unobserved particle in

Fig. 3.3. The strangeness versus isospin plot for the family of resonances of spin $\frac{3}{2}$. When this was first plotted only the top nine particles were known. It is obvious from symmetry that a tenth particle should occupy the lowest apex of the triangle. This particle, the Ω^-, was later discovered.

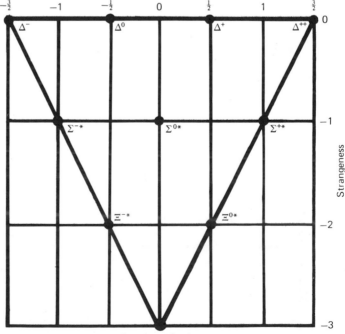

considerable detail. With the sort of lifetime and mass they predicted the particle should be both producible and observable in a bubble chamber at the then largest accelerators. A search was made for it and, in 1964, it was found. Figure 3.4 shows the bubble chamber photograph found by V. E. Barnes and 32 others at Brookhaven National Laboratory, USA. They used a beam of K^- mesons of energy 5 GeV. This beam was fired into their 80-inch bubble chamber.

Bubble chambers, like Wilson cloud chambers, are devices for making visible the tracks of fast charged particles. They were invented by Donald Glaser of Chicago for which invention he was awarded the Nobel Prize in 1960. In the chamber, a liquid is kept at a temperature just below its boiling point by application of pressure. If the pressure is momentarily released it is then at a temperature just *above* its new boiling point. If charged particles then pass through the chamber boiling starts on the ions produced by the particle's passage – so the path of the particle is delineated by a line of bubbles. A photograph is quickly taken and the pressure re-applied. With accelerators, one can arrange to release the pressure just before the beam enters the chamber, so these are very useful devices to use with accelerators.

Originally, bubble chambers were made with liquids boiling around room temperature. But hydrogen is often the best target material for studying particle collisions. The target nucleus is then just one proton, so the collision is as simple as one can manage and not complicated by cascade processes that may occur in a more complex nucleus. But hydrogen boils at a very low temperature, 20 °K or 253 °C below freezing point. Also hydrogen bubble chambers were difficult to make. They were developed by Luis Alvarez of Berkeley, who in 1968 was awarded the Nobel Prize for this and other work.

Luis Alvarez is a very ingenious person. He used cosmic radiation to X-ray the Great Pyramid, looking for unknown interior chambers (there were none). Since his 'retirement' he, his son, and other workers have put forward a very credible theory of the sudden extinction of the dinosaurs and produced a good deal of evidence to substantiate it.

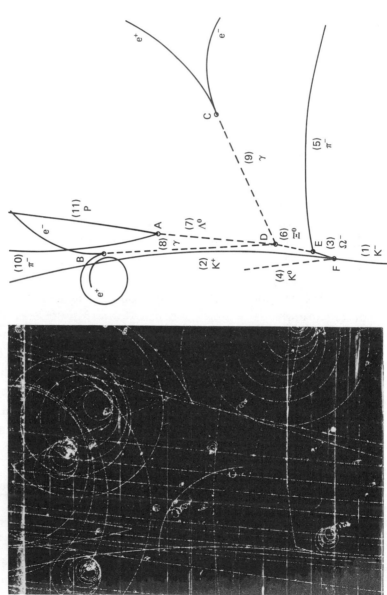

Fig. 3.4. The bubble chamber picture that established the existence of the Ω^- particle. The incident particle that produced the interaction was a negative kaon (K^-, at the bottom of the picture). The complex decay chain after production is given in the line diagram and explained in the text. The photograph and diagram were generously provided by the Ω^- Group of the Brookhaven National Laboratory.

Barnes *et al.* fired a beam of negative kaons into this liquid-hydrogen-filled bubble chamber. After about 100 000 photographs, they got one that showed the production of an omega minus (Ω^-) particle (fig. 3.4). The production was followed by a whole series of decays, like this:

$$K^- + p^+ \to \Omega^- + k^+ + k^0$$
$$\searrow$$
$$\Xi^0 + \pi^-$$
$$\searrow$$
$$\Lambda^0 + \pi^0$$
$$\searrow$$
$$\pi^- + p^+.$$

In addition to this main chain of decays one also saw the decay of $\pi^\pm$ and π^0, k^+, k^0 and so on (or if one did not see them, they happened outside of the chamber). The line drawing accompanying the photograph in fig. 3.4 shows the way in which the various charged tracks arose.

The bubble chamber was, as is usual, in a strong magnetic field. (Nowadays it is usual to use superconducting magnets for this – it saves electricity and means there is no large quantity of ohmic heat to get rid of.) The curvature of the charged tracks in this magnetic field allows us to calculate their momenta and, using the conservation of energy and momentum, we can do a 'book balancing' calculation in which the only thing we do not know initially is the mass of the Ω^-. So Barnes *et al.* determined that mass. It came out as 1686 ± 12 MeV – very close indeed to the predicted value.

Thus the application of these ideas about symmetry led to the prediction and discovery of this new and remarkable (strangeness $= -3$!) particle. Not only a very 'strange' particle, in the technical sense, but unusual in that it is a 'long-lived' completion of a very short-lived family! (What use it is, in the Universe as a whole, is another question and, so far, more difficult to answer.) But does anything lie behind these symmetries? Is there some simplifying principle?

In 1963, two people suggested that there was. One was

Murray Gell-Mann, the other George Zweig. Independently (working at two ends of the same corridor in Caltech, I have been told) they proposed that all the many hadrons, baryons and mesons alike, are made from just three 'really fundamental' particles (and their anti-particles). Well, three according to Murray Gell-Mann, who called them 'quarks' after a quotation from James Joyce's novel, *Finnegan's Wake*: 'Three quarks for Muster Mark, Sure he hasn't got much of a bark, etc.'. (It is worth noting that Joyce rhymes quark with bark, Mark, lark, etc.) George Zweig may have been somewhat more intuitive – he called them 'aces', and it is well known that there are four aces in most packs. However, for the particles known in 1963, three were enough and the name 'quark' has stuck.

The scheme supposed that all baryons are made of three quarks. The quarks have been given the identifying letters u, d and s. These are sometimes taken to stand for up, down and sideways (or, maybe, strange). A proton is supposed to consist of two up quarks and one down quark, so we could write a proton as uud. Now, how about the characteristics of a proton? First of all, its electric charge is $+1$ and because electric charge is always conserved the total charge of two up quarks and one down quark must be $+1$. So we can write $2Q_u + Q_d = +1$. A neutron is supposed to consist of two down quarks and an up quark so $Q_u + 2Q_d = 0$. We have two simple equations and the solutions are $Q_u = \frac{2}{3}$ and $Q_d = -\frac{1}{3}$. That was a bit of a shock. The quarks are fractionally charged and the reaction of most physicists was 'impossible – I mean, unlikely – no one has ever found a fractional charge!'

However, there was a further surprise when the records were checked. Robert Millikan was the first man to measure the charge on the electron (he was awarded the Nobel Prize for it in 1923) and he had written in his paper of 1910 in the *Philosophical Magazine*

> I have discarded one uncertain and unduplicated observation apparently upon a singly-charged drop which gave a value of the charge some 30 per cent lower than the final value of e.

That is, Millikan found one charge very close to $\frac{2}{3}$e (e is the charge on the electron) and, although he did not believe it was real, he reported it. That is very good experimental procedure. However, since then very few people had reported any sort of fractional charge, let alone precise values of $\frac{2}{3}$ or $\frac{1}{3}$e. So these suggested values for the quark charges made them unusual particles, and also suggested that they should be easy to distinguish from 'ordinary' integrally-charged particles.

They should, for instance, leave very distinctive tracks in cloud or bubble chambers, or photographic emulsion, and give obviously different signals when they passed through scintillation counters. For fast particles, ionisation is proportional to the square of their charge. With a charge of $+2$, a fast α particle ionises four times as much as a fast proton (and a fast uranium nucleus with a charge of 92 ionises 8464 times as much as a proton). So these two quarks should only ionise $\frac{4}{9}$ or $\frac{1}{9}$ as much as a fast proton, or electron. This alone should make them easy to detect. Or, if the quark is at rest, then it might be possible to detect it in the sort of experiment Millikan was doing when he determined the charge on an electron and found his one anomalous charge.

However, before I consider in detail how to hunt for quarks, I will take a look at some of their other properties and at how we can make all the various particles from them.

First their spin. Three quarks are needed to make a proton, which itself has spin $\frac{1}{2}$; so the spin of the quark must be half integral and most likely is just $\frac{1}{2}$. Then, their baryon number must add to 1, which is the baryon number for a proton. None of the three quarks in a proton is an anti-quark so all three baryon numbers are positive, so their baryon number is $\frac{1}{3}$, fractional again, like their electric charge.

In a similar way we can determine their isotopic spin (I) and its third component (I_3) and their strangeness (s). The up and down quarks, out of which protons and neutrons are thought to be made, have zero strangeness, whilst the s quark has a strangeness of -1.

Their masses are not so easy to determine. Obviously the mass of the three quarks making a proton must be greater than the proton mass. But when particles bind together to form a more complex system they use part of their mass energy to do the binding. A deuteron is made up of a proton and a neutron. The sum of the proton and neutron masses is (from table 3.1) 1877.9 MeV. The deuteron mass is 1875.7 MeV, so the neutron and proton have used 2.2 MeV to bind themselves together. The reverse way of looking at this is to say that it will take an energy of 2.2 MeV to separate a deuteron into its component neutron and proton. Now the quarks in a proton are bound together by some force as yet unknown – it may be a *very* strong force. If so, the binding energy will be large and could be much greater than the mass of the proton. So quarks may be very heavy particles. In any case they must be more massive than about 330 MeV, roughly one-third of the mass of the proton.

So we can make a table of the properties of the three quarks (table 3.2).

Now it becomes possible to play a game of Make the Particle. The rules are simple: baryons are to be made out of three quarks (anti-baryons out of three anti-quarks) and mesons are to be made from a quark–anti-quark pair. Anti-quarks have quantum numbers of the same magnitudes as quarks but of the opposite sign. Hence using a quark–anti-quark pair means we get a baryon number of zero – which means the particle is a meson. To see how this works consider the Λ^0 particle. It has charge zero, strangeness -1, spin $\frac{1}{2}$, baryon number 1, and $I_3 = 0$. So we make it out of a u, a d and an s quark, as in table 3.3. You might think that I have cheated with the spin column and

Table 3.2 *The properties of quarks.*

Quark	Charge in units of e	Spin in units of $\hbar$	Baryon number	Strange-ness s	Isospin I	I_3
u	$\frac{2}{3}$	$\frac{1}{2}$	$\frac{1}{3}$	0	$\frac{1}{2}$	$\frac{1}{2}$
d	$-\frac{1}{3}$	$\frac{1}{2}$	$\frac{1}{3}$	0	$\frac{1}{2}$	$-\frac{1}{2}$
s	$-\frac{1}{3}$	$\frac{1}{2}$	$\frac{1}{3}$	-1	0	0

that three lots of $\frac{1}{2}$ should make $\frac{3}{2}$. But remember that the spins are vectors – all I have done is have two $\frac{1}{2}$ spins pointing up and one pointing down, to give a total spin of $\frac{1}{2}$. The totals for all the properties agree perfectly with the properties of the Λ^0 baryon.

Now the Λ^0 is a member of the family shown in table 3.4. This family all have spin J of $\frac{1}{2}$ and what is called the parity of their wave functions is positive (that is, the wave function does not change a sign when we apply the P operator). So the family is called the $J^P = \frac{1}{2}^+$ family. The quark scheme for this family is shown in table 3.4. We can next move on to the $J^P = \frac{3}{2}^+$ family, which is the family that includes the Ω^- particle. The quark scheme for this is shown in table 3.5. Once again, we see that each particle can be explained as a combination of three quarks.

Table 3.3

Quark	Charge in units of e	Spin in units of $\hbar$	Baryon number	Strangeness s	I_3
u	$\frac{2}{3}$	$\frac{1}{2}$	$\frac{1}{3}$	0	$\frac{1}{2}$
d	$-\frac{1}{3}$	$\frac{1}{2}$	$\frac{1}{3}$	0	$-\frac{1}{2}$
s	$-\frac{1}{3}$	$\frac{1}{2}$	$\frac{1}{3}$	-1	0
Total	0	$\frac{1}{2}$	1	-1	0

Table 3.4 *The constitution of the particles in fig. 3.1 suggested by the quark model. The proton and neutron are two members of the $J^P = \frac{1}{2}^+$ family.*

Particle	p	n	Λ^0	Σ^+	Σ^0	Σ^-	Ξ^0	Ξ^-
Quarks	*uud*	*udd*	*uds*	*uus*	*uds*	*dds*	*uss*	*dss*

Table 3.5 *The constitution of the particles in the $J^P = \frac{3}{2}^+$ family suggested by the quark model.*

Particle	Δ^{++}	Δ^+	Δ^0	Δ^-	Σ^{+*}	Σ^{0*}	Σ^{-*}	Ξ^{0*}	Ξ^{-*}	Ω^-
Quarks	*uuu*	*uud*	*udd*	*ddd*	*uus*	*uds*	*dds*	*uss*	*dss*	*sss*

The Ω^-, with its strangeness of -3 and single negative charge, comes out naturally as the sum of three s quarks.

Next the mesons. The positive pion is the sum of u quark and a d anti-quark. This gives charge of $+1$, spin, baryon number and strangeness of zero – just what is required. And so we can go through all the family of mesons and baryons, including all the short-lived particles not mentioned individually in table 3.1. All the particles can be made in this simple way, and all the particles that can be made in this way have been found. This is good evidence for correctness of the hypothesis.

But there are difficulties. For instance, the quarks are fermions; they have spin $\frac{1}{2}$ so they should obey the Pauli Exclusion Principle. But, in Ω^- for instance, we have three s quarks all in the same state. There seemed to be only two ways out of this. One is to say that the Pauli Exclusion Principle breaks down for quarks. The other suggests that not all s quarks are alike. They have perhaps some extra property that differentiates them one from the other, just as some property differentiates 'electron neutrinos' from 'muon neutrinos'. This latter seemed to be much the more preferable of the two suggestions. So it was assumed all quarks come in three 'colours'. There was some discussion about the colours; eventually, primary colours red, green and blue were chosen. The anti-quarks can then have the opposing colours on the colour wheel, cyan, magenta and yellow. It became clear that in nature, only 'white' combinations of quarks occur as known particles. It is rather surprising how well this colour analogy applies.

But, of course, we have disposed of the difficulties of the Exclusion Principle only by increasing the number of fundamental building blocks again. Now we need *nine* quarks, three 'flavours' (as u, d and s have come to be called), each coming in three colours, red, green or blue (and, of course, their anti-quarks).

Another difficulty at the time was that (with the possible exception of Millikan) none one had seen a free quark. Accelerators had reached energies of 30 GeV – so collisions of 30 GeV protons on stationary protons were very frequent indeed. But no quark had been obvious. So began a great hunt for quarks. There were three obvious places to look: at the big accelerators; in cosmic radiation; and in solid (or liquid) matter. I will now discuss the methods one can use in this research and then, in the next two chapters, review the searches for free quarks and various other experiments bearing on the question of their existence.

Fractional charge is the most obvious property that distinguishes quarks from particles like the electron, muon, pion, or proton that have a single charge. If the quark is at rest in matter then it should be possible to carry out a Millikan-style experiment to measure the fractional charge directly. If the quark is moving quickly (that is, with a speed approaching that of light) then one needs to measure its ionisation. I will discuss the ways of doing this first.

Suppose, to begin with, that the particle is moving through a gas. It encounters many atoms. If the encounter is fairly close, the particle may detach one of the outer electrons from the atom. It may only just manage to do this; or it may, if the encounter is closer, give the electron quite a considerable amount of energy. In any case, the first result is the formation of an ion pair – the electron (called the negative ion) and the atom, now minus one of its electrons (called the positive ion). In going through one centimetre of air at ordinary temperature and pressure, a fast, singly-charged particle will make, on average, between 40 and 50 of these pairs. If the electron is hit with enough energy it can itself produce more ions and this adds somewhat to the total ionisation produced by the original fast particle.

In 1899, C. T. R. Wilson at Cambridge found a way of measuring this ionisation. He found that if moist, dust-free air is expanded and suddenly cooled, becoming supersaturated with water vapour, water droplets form preferentially on any *ions* present. If the supersaturation is too great, drops will form anywhere; if too low, they do not form at all. But within a reasonably wide range of supersaturation, the droplets form on the ions and grow big enough to photograph and then to count. The device is called a cloud chamber; it not only enables us to measure the number of ions per unit length of the track but it also makes the tracks visible. As we have already seen, it has been a very useful tool in the study of subatomic particles.

Since C. T. R. Wilson's original apparatus, many improvements have been made. Figure 3.5 shows a modern cloud chamber. It is divided into a front and back chamber by a neoprene diaphragm, which can move between two perforated

Fig. 3.5. This shows the plan of a Wilson cloud chamber. The chamber has been somewhat dissected so that the front chamber, back plate and back chamber are shown separated. In operation they are all clamped together so that both back and front chambers are gas tight and separated from each other by the neoprene diaphragm.

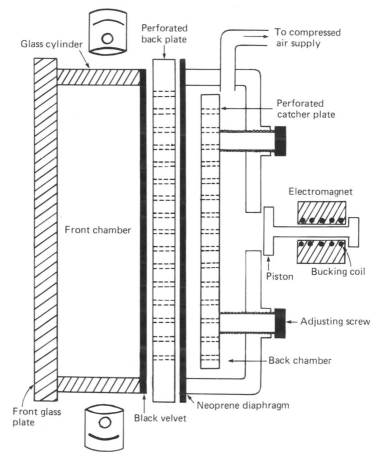

metal plates. One of these forms the back wall of the front chamber. The other is called the 'catcher' plate. Its position can be adjusted by means of three screws. There are two openings in the back chamber. One is closed by a piston valve, which is normally held shut by an electromagnet. The second is connected to a supply of compressed air of regulated pressure. This supply is adjusted to be somewhat higher than the pressure in the front chamber, which itself is somewhat higher ($\sim 1\frac{1}{3} \times$) than atmospheric pressure. So the neoprene diaphragm is normally pressed forward against the back of the front chamber. The gas in the front chamber is kept saturated with a mixture of alcohol and water vapour by keeping a small open glass vessel of this mixture in the bottom of the chamber. To expand the chamber we pass a momentarily large electric current through the 'bucking coil' of the magnetic valve. This current is in the opposite direction to the current in the main holding coil and momentarily collapses the magnetic field. The pressure in the back chamber then blows off the piston. The air rushes out of the large opening (with a bang) and the pressure in the back chamber drops to normal atmospheric pressure in a few milliseconds. The pressure in the front chamber forces the neoprene diaphragm back onto the catcher plate. This produces the desired degree of supersaturation in the front chamber; droplets condense on all the ions present and the tracks of any charged particles are made visible.

After an expansion the chamber is 'sensitive' for a short time – depending on the size of the chamber. The droplets grow and can be photographed about 0.1 s after the expansion. Before that they are too small to photograph well; after that, convective movement of the gas inside the chamber starts to distort the tracks. The light for photography is generally provided by xenon-filled flash tubes, which give a very short, very bright pulse of light. Two of these are shown on either side of the chamber in fig. 3.5. They are arranged to provide a parallel beam of light across the chamber. The photographs (very often stereo photographs) are taken through the front wall of the front chamber, which is made of high quality plate glass. The front of the back plate is covered with black velvet to make a dark background against which the tracks stand out clearly.

After the rapid expansion, the chamber is generally expanded slowly two or three times to settle any dust stirred up by the bang. Then it is recompressed, the film wound on, and it is ready for the next expansion. All the various procedures are carried out at the appropriate times by electronic control circuitry. The initial expansion is also initiated by a pulse from the event selection system, which in cosmic ray work will be a set of some sort of particle detectors arranged to select events likely to be of interest in the particular experiment. And this is one sort of trap that one can set for quarks.

Another detector, similar in principle to the cloud chamber, is the bubble chamber. In the cloud chamber, the ions are the nuclei on which drops of mist (or cloud) form in a supersaturated vapour. In a bubble chamber they are the nuclei on which bubbles of vapour form in a superheated liquid. But a liquid is much denser than a vapour and things go more quickly. The ions recombine much faster – so fast that it is not possible to have the ionising particle pass through the system first and then lower its pressure to make it superheated. Before this can be done the ions have recombined. So a triggered bubble chamber is not feasible and bubble chambers are not used in cosmic ray experiments. However, with accelerators one knows just when the fast particles will arrive so the bubble chamber can be expanded just before the beam arrives. Then inject the beam, wait a very short time, photograph the bubbles, and whack the pressure back on fast to stop the liquid from boiling explosively. It sounds a desperate sort of game, especially when one realises that the chambers are *big*, maybe 8 m long, and the liquid, as often as not, is liquid hydrogen, at ~ -240 °C. But with good modern engineering, they work and produce very beautiful pictures of these tracks from high energy particle interactions, for example fig. 3.4. There are other elegant examples in Fritjof Capra's book *The Tao of Physics*.

Photographic emulsions, of course, also make particle tracks visible. But they are continuously sensitive unlike cloud and bubble chambers, which are sensitive only for brief periods just after expansion. For many purposes this continuous sensitivity is a great advantage. But not for quark hunting. Quarks are expected to be quite rare – otherwise experiments in the past, like Millikan's oil drop experiment, would have seen numbers of them. So one wants a device that will trigger only when some event presents the right sort of 'signature'. Another difficulty is that a track with only $\frac{1}{9}$ of the ionisation of a fast proton would be very difficult to spot in an emulsion – that ionisation would be only between 2 and 3 developed grains every 100 μm. So emulsions would only be useful for detecting the quarks with charge $\frac{2}{3}$e.

Another 'visual' device is a bank of neon flash tubes. The neon flash tube is a very simple device – just a long (maybe 2 m) glass tube, 1 or 2 cm in internal diameter and filled with neon. If such a tube is placed between two electrodes it will glow when a high voltage pulse is applied to the electrodes within a few microseconds after a charged particle has passed through the tube. The glow discharge is started by the ionisation produced by the particle. If one has a bank of these tubes, maybe two dozen deep and wide, and photographs it from the end, a two-dimensional projection is obtained of the path of any charged particle passing through the bank just before the picture is taken. The tubes can be made to be about 98 per cent efficient for normal singly-charged fast particles. But because of the smaller ionisation produced by quarks ($\frac{4}{9}$ or $\frac{1}{9}$ that of singly-charged particles), they are appreciably less efficient at detecting quarks. This can be used to single out any quarks hitting the bank.

The neon tube detector was invented by Marcello Conversi. It led on to another device that enables the trajectory of a particle in space to be determined. This is the spark chamber. A spark chamber consists of a pair of parallel electrodes in a suitable gas at a suitable pressure. If a high voltage pulse is applied across the electrodes just after a particle has passed through them (and the gas), a spark will jump between them, following the ionised column of gas left behind by the particle.

The position of the spark can be determined by photographing the gap from two directions (maybe at right angles) or by various other ways. If we use a number of these chambers suitably dispersed we can determine the trajectory of the particle. Spark chambers are not useful in determining the ionisation produced by a particle. They have been used in quark searches to determine the particle trajectory.

The next instrument I will describe does not make the particle trajectory visible but does make a measure of the ionisation. This device is called a scintillator, for short. It is a combination of a block of scintillating material and an electronic device, called a photomultiplier, which converts weak flashes of light into electrical pulses and greatly amplifies the pulse. Suitably treated, a photomultiplier can count individual photons. The scintillator is the modernised version of the scintillating screen used in the early days of radioactivity to detect individual α-particles. In these experiments the photon detector was a human eye which, in some cases, can also detect individual photons.

In the modern instrument, the scintillating material is very often a block of transparent plastic containing two organic chemicals. The first of these is often paraterphenyl – in any case, it is a chemical whose molecules emit a photon when excited by the passage of a fast charged particle. The second chemical is more complicated and has a longer name, which is generally replaced by the acronym POPOP. This chemical is called a 'wavelength shifter'. It is necessary because many of the photons emitted by the paraterphenyl are in the ultraviolet region of the spectrum, whereas the photomultiplier responds better to photons in the visible region. So the POPOP takes in photons in the ultraviolet region and re-emits them in the visible region. The scintillating material is made by dissolving appropriate amounts of the materials in the 'monomer', the liquid which is to form the plastic. This liquid is then encouraged to polymerise and turn into a transparent solid. This can be done in a container (or mould) of the shape required.

The photomultiplier has a glass window and, backing that, a very thin layer of metal that emits electrons when hit by the photons. Behind this photocathode, as it is called, is a series of electrodes (called dynodes) each at a progressively higher positive potential. The electrons from the photocathode are attracted to the first dynode and accelerated by the potential difference between them. When they strike the first dynode they have enough energy to knock out several electrons each. These are then attracted to the second dynode and the process repeated. The art of designing a photomultiplier is to ensure this multiplying cascade goes smoothly. Very large amplifications (maybe of a million times) can be achieved. The final pulse of electrons is deposited on the *anode* and fed into the electronic circuitry.

The combination of scintillator and photomultiplier makes a very fast and versatile particle counter. Times are measured in nanoseconds (10^{-9} s) rather than microseconds. The device is also able to distinguish between particles of different ionisations. For instance, if we have a scintillator–photomultiplier combination that normally gives 100 electrons out of the photocathode for the passage of a fast proton through the scintillator, then it will normally give $\frac{4}{9} \times 100 \simeq 44$ electrons for the passage of $\frac{2}{3}e$ quark and 11 electrons for a $\frac{1}{3}e$ quark. However, if a proton normally gives 100 electrons out of the cathode it will, because of statistical fluctuations, give 90 or fewer electrons about 15 per cent of the time and it has a small but noticeable chance of only giving 70 electrons. Similarly, the $\frac{2}{3}e$ quark may sometimes give greater than 60 electrons with about the same probability. If the mean number of electrons is smaller, then the statistical fluctuations are proportionally bigger and we can get an appreciable overlap in the two distributions. So for good identification we generally need measurements in several scintillators in succession.

These are the principal devices used in the search for fast quarks. To search for quarks that may be at rest in matter, people have used a number of techniques. The principal devices have mostly been modernised versions of Millikan's method of determining the charge on the electron (and for showing that there is a unique, smallest possible electric charge).

What Millikan did was to introduce a mist of very fine

droplets of oil or water between two parallel horizontal metal plates. Then he irradiated the mist with ultraviolet light, which knocked one or more electrons from some drops. The electrons were picked up by other drops and he then had a mist with some charged drops. The drops, of course, were all the time falling under the Earth's gravitational field. He then applied an electric potential to the plates to counter-balance the Earth's pull for one particular drop that he had selected for observation in a low-powered microscope looking in to the mist. The electrical field was adjusted to get exact balance and then, after measuring the size of the drop, Millikan was able to calculate the electric charge on the drop just by equating the two forces, the downward pull due to gravity and the upward pull due to the electrostatic attraction, He found, with the one exception I have noted already, that all the measured charges were integral multiples of a certain very small charge and this he gave as the charge on the electron. He was awarded the Nobel Prize in 1923 for this work.

Now the droplets in Millikan's experiment were tiny. They only weighed about 10^{-14} kg. When people thought of searching for quarks in matter they expected them to be rather rare and so wanted to look at much larger amounts of matter. So they looked for ways of doing Millikan's experiment with much bigger objects than the droplets of mist. In chapter 5 I describe some of these elegant experiments.

These then are the main ways in which people have looked for quarks. Some of them are very expensive – those involving large accelerators for instance. But others are fairly simple, cheap experiments. Some, in fact, could be built and operated by a group of amateur scientists of the standard of the amateur astronomers who construct such admirable large telescopes and organise such projects as meteor watching.

4

The great accelerator hunt

When Gell-Mann and Zweig put forward their quark hypothesis, the maximum energy available at accelerators was 30 GeV (at Brookhaven National Laboratory, New York). This machine accelerated protons to this energy and then collided them against a stationary target. Machines of this type have now (May 1982) reached energies of 500 GeV (NAL) and 300 GeV (CERN). NAL is the National Accelerator Laboratory (often called Fermilab) at Batavia, Illinois and CERN is the Centre Européenne pour la Recherche Nucleaire at Geneva, Switzerland. But there is a class of accelerator, called colliding beam accelerators, in which, instead of colliding an accelerated beam of particles with a stationary target, we collide a beam with a beam. The first large machine of this type is the Intersecting Storage Rings (ISR) device at CERN. It collides two beams of 30 GeV protons. Because of relativity effects this is equivalent to colliding a beam of $\sim$ 2000 GeV protons with a stationary target. Recently, CERN has brought its SPS device into operation. This collides two beams, one of protons, the other of anti-protons, at energies of 250 GeV each. This is equivalent to an energy rather greater than 100000 GeV in a machine with a stationary target.

In 1963, 30 GeV seemed a large energy – quite enough to liberate a quark from a proton. One method of checking the quark hypothesis was to collide the protons and search the debris for fractionally-charged particles, with their characteristic 'low ionisation' signature. Another method was to examine the protons, looking for their internal structure, in the same way that Rutherford and his co-workers had looked at the structure of atoms. In his case he used α-particles as the probe particle and discovered the atomic nucleus. The modern method is to use electrons or other protons as the probe, hoping to find the much smaller quarks inside the proton. A third method is to study new particles produced in these high-energy collisions to see if they could all be fitted into the quark scheme.

I shall begin with the first of these methods, the search for free quarks at accelerators. The first search was made in 1964 – very soon after quarks were postulated. Scientists at the biggest accelerators already had many photographs of interactions of high-energy protons in bubble chambers. They began to scan these films for signs of particles with low ionisation. The accelerators used were the CERN proton synchrotron, giving an energy of 25 GeV, and the Brookhaven synchrotron, giving 31 GeV. All groups looked for quarks of charge both $\frac{1}{3}$e and $\frac{2}{3}$e. None were found. This could mean that free quarks do not occur; or that quarks do not exist; or that if they exist their mass is greater than $\sim$ 2.5 GeV. This last is because quarks of energy greater than 2.5 GeV could not be produced with the energy available.

When higher energies became available the searches were repeated. Searches were made, using protons as the bombarding particle, at energies up to 300 GeV and still no free quarks were found.

Then the CERN intersecting storage rings became available. This is the accelerator which 'clashes' together two beams of protons, moving in opposite directions. Two searches have been made at this accelerator at energies equivalent to 1500 to 2000 GeV in a fixed target arrangement. Figure 4.1 shows a diagram of one of these experiments carried out by Fabjan and nine other scientists (including Laurence Peak of my Department in Sydney). It is a much simplified diagram, which may give you some idea of the elaboration, size and detail of experiments carried out at the large accelerators these days. The two evacuated tubes carrying the proton beams are labelled Ring 1 and Ring 2. The circumference of these rings is almost 1 km. The protons are bent into the roughly circular path by bending magnets along this circumference. The beams clash, not quite head on, at the intersection of these two rings. This is where the interactions of interest take place.

A number of 'telescopes' are placed to detect any quarks produced in the high energy interactions. The 'telescopes' contain scintillation counters, which I have described in chapter 3; multiwire proportional counters, which are a modification of Geiger counters; and Cerenkov counters. The proportional counters are used to determine the positions of the particles as they cross various planes in the telescope, and hence to allow us to calculate the trajectory of the particle. In this experiment all the trajectories should be straight lines because no magnetic fields were used in the telescopes. Cerenkov counters are named after P. A. Cerenkov, who discovered the electromagnetic shock wave effect on which the counters operate. He was awarded the Nobel Physics Prize in 1958. When an object, such as a high-speed airplane, moves faster than the speed of sound in air, it produces a shock wave. Now the speed of light in transparent materials such as glass, perspex or lucite, or air, is less than the speed of light in vacuum. No material object can exceed (or even reach) the speed of light in a vacuum. But they can exceed the speed of light in such materials and when they do they produce an electromagnetic shock wave, a cone of light which can be picked up by visual detectors – photomultipliers or our eyes, for

Fig. 4.1. A diagram of the quark search experiment carried out by Fabjan *et al.* at the Intersecting Storage Rings of CERN, Geneva.

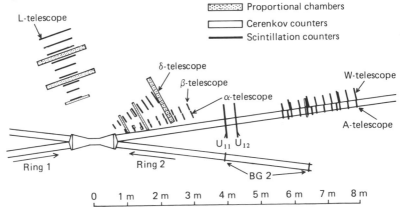

instance. This enables us to determine limits on the speed of particles by getting them to pass through materials in which the speed of light is known and seeing whether or not they produce Cerenkov radiation. Such devices are called Cerenkov counters.

With this device Fabjan and his co-workers looked at something like 4 million events! They found *one* particle that had all the characteristics expected of a quark. It passed all the tests that were supposed to reject 'false quarks'. But with only one event, and the possibility that despite all their precautions the experimenters had overlooked some possibility of 'fraud', they were loath to claim it as evidence for the existence of free quarks.

The second experiment at the ISR was carried out by a team led by Antonio Zichichi. His experiment improved on the Fabjan experiment by including a very large analysing magnet. This enabled the momentum of the particles to be determined and this, combined with a measurement of the ionisation, allowed their mass to be calculated. 150000 candidates were considered and all but one rejected. This one passed all the tests for a quark: its measured charge was $\frac{1}{3}$e; its velocity 0.96 that of light, and its mass 169 MeV. Again, because of the possibility of some unknown process faking a quark signature, they were reluctant to claim it as a genuine quark track. Another point is that the measured mass, 169 MeV, is uncomfortably light.

So direct searches for free quarks at the large accelerators have not led to any certainty about the existence of quarks. However, the other two methods I mentioned at the beginning of this chapter have had a lot more success (in finding quarks, that is, *not* in finding free quarks). Of these two, the first that I mentioned was the updated version of the Rutherford scattering experiment. Rutherford first noted the scattering of α-particles in 1906, when he was endeavouring to measure their deflection in a strong magnetic field. Generally, he carried out these attempts in a 'good vacuum'. But in a few tests, in which he allowed in a small amount of air, he noted that the paths of some of the particles were somewhat bent. Using a very thin mica sheet (without any air) produced the same result. It was an apparently trivial result, which many people would not have followed up. But Rutherford's intuition in this field was remarkable. He realised that no deflecting agency he could set up (such as a magnetic or electric field) could produce a deflection of one or two degrees in a distance equal to the thickness of the mica (~ 0.003 cm). Over that distance he would need an electric field of ~ 100 million volts per cm! He set two of his best people to work on the problem; Hans Geiger (who invented the geiger counter) and Marsden. Within a few months they had found some deflections of more than 90°. By 1910 Rutherford realised that this must mean that the atom has a small positively charged heavy nucleus. As I have noted earlier, it is *very* small compared with an atom; one hundred thousand times smaller. So Rutherford had probed the structure of the comparatively large (diameter 10^{-10} m) atom using fast α-particles. He did not know when he started that the α-particles themselves are tiny. They are nuclei of helium atoms and their diameter is $\sim 10^{-15}$ m. So what Rutherford was doing was probing the atom using a beam of these tiny entities as one might use a steel spike to probe soft mud for a hidden treasure chest. And in the 1960s physicists again turned to this method to probe the proton.

This time, the beam was a beam of very high energy electrons (~ 20 GeV $\equiv 2.10^{10}$ electron volts). There are two points to consider here: one is the nature of the probing particle and the other is its energy. As far as we have been able to measure, electrons are 'point particles'. We have been able to measure down to $\sim 10^{-18}$ m, that is to about $\frac{1}{1000}$ the radius of a proton (or an α-particle). So the modern probe is a much finer instrument than Rutherford's α-particles. Secondly, looking at it from the wave point of view (rather than as particles), we know from de Broglie that the higher the momentum the smaller the wavelength. The energy of the electrons in the new experiments is ~ 10000 times the energy of Rutherford's α-particles; so their momentum is about 5000 times that of the α-particles and hence their wavelength is shorter by just that factor. So on both counts we have a very fine probe. And, of course, we may need it. In

Rutherford's experiment the 'treasure', the atomic nucleus, was very much smaller than the atom and one expects the same to be the case for the quarks in the proton.

The instrument that provided the electrons and first carried out this sort of experiment was the Linear Accelerator at Stanford University, California. As its name says, this is not a circular machine, but straight – two miles long. Electrons are made to 'surf ride' on an electromagnetic wave down the two-mile long vacuum pipe to reach this very high energy. They then collide with the protons and their scattering angles are measured. The results were a great vindication of the quark theory of the structure of the proton. Three point-like scattering centres were found inside the proton, carrying the fractional electric changes predicted by quark theory; two centres with charge $\frac{2}{3}$e and one with charge $-\frac{1}{3}$e. Their spins were measured and again agreed with the prediction.

The third method also provided good evidence for the truth of the quark hypothesis. In fact, it went further. It not only justified the hypothesis; it extended it. This third method, remember, was to search for new, previously unknown particles and to see if they fitted into the quark scheme. Many particles were found that did indeed fit into the quark scheme. For a long time none were found that failed to fit. And when at last a particle appeared that could not be placed in the then-existing quark scheme, it was fairly obvious that it (and other particles like it) could fit an extended scheme that included more than three flavours of quark.

The first hint of this new type of quark came from experiments in cosmic radiation. K. Niu, E. Mikumo and Yasuko Maeda, working in the University of Tokyo's Institute for Nuclear Studies and Yokohama National University, used a device called an emulsion chamber to study high energy interactions of the cosmic radiation. The emulsion chamber consisted of 79 layers of nuclear research emulsion interspersed with layers of lead in the top and bottom sections, and with layers of acrylic in the middle section. A chamber was quite small, $0.2 \times 0.25 \times 0.19$ m³. Twelve of them were flown in a JAL cargo jet for about 500 hours at about 10000 m altitude. At this altitude they only have about a quarter of the atmosphere above them so the cosmic radiation is, by sea level standards, intense, and high energy interactions fairly frequent. One of these high energy events produced 70 fast, singly-charged particles and 19 slow, heavy fragments. The 19 slow fragments immediately imply that the target nucleus was that of silver or bromine, because these two are the only elements in emulsion with more than 19 nucleons in their nucleus. The projectile was a neutral particle and so, most likely, was a neutron, although a neutral kaon is a faint possibility. This follows from the lifetimes of the neutral particles. Two of the fast singly-charged tracks showed sudden bends like the charged track in the second of the Rochester–Butler cloud chamber photographs (fig. 2.2). Associated with these kinks were two neutral pions, which became obvious when the γ-rays from their decay produced large showers of electrons and positrons deeper in the chamber.

Figure 4.2 shows a sketch of the event. The size of the electromagnetic showers allows an estimate of the energy of the neutral pions to be made. In each case, the plane defined by the track of the charged particle after the kink and the path of the neutral pion contains the path of the track before the kink. Niu could then argue that this means that the decay is into only two particles (because there is no unbalanced component of momentum). And then, if he assumed the charged particle also to be a pion, he could calculate the mass and lifetime of the new particle (which he called an X-particle). He got a mass of 1.78 GeV (almost twice that of the proton) and a lifetime of $\sim 2 \times 10^{-14}$ s. If he assumed the charged particle after decay to be a proton the mass was even heavier. Very quickly Dr S. Ogawa pointed out that such a heavy particle had almost no chance of living as long as 10^{-14} s unless there was some new quantum number involved (just as had been the case with the Rochester–Butler particles). This in turn, if we wish to retain quark theory, meant that there must be a fourth flavour of quark; a quark carrying the new quantum number, which eventually became known as 'charm', maybe because the new particles seemed to bear a charmed life.

This work was published in 1970 and in the next few years

other groups working in cosmic radiation confirmed the existence of the heavy, comparatively long-lived particles. Then in 1974 two groups, both in the USA, discovered the existence in accelerator events of somewhat similar particles. The particular particle they found was comparatively long lived – $\sim 10^{-20}$ s. It does not sound long, but it is 1000 times longer than it would have been if the particle could decay via the strong interaction.

One of the two simultaneous experiments was carried out at the Brookhaven National Laboratory on Long Island, New York. The group was led by Samuel Ting. The other experiment

Fig. 4.2. A diagram of the type of event found by Professor Niu and his collaborators. A high energy interaction takes place at the top of the diagram (X). The tracks of a few of the many pions emerging are shown as straight black lines. Two of the tracks, however, are charmed particles, which decay after only a short distance producing characteristic kinks in the track, and neutral pions, which leave no track but whose direction is made visible by the reactions of their own decay γ-rays. The directions of the neutral pions are shown as dotted lines.

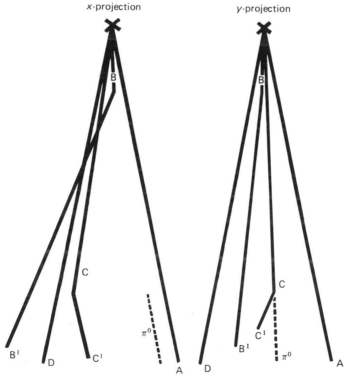

was carried out at the far side of the USA, at the Stanford Linear Accelerator in California. I will describe this latter experiment because that type of accelerator turned out to be a very useful device for investigating the charmed particles.

The Stanford Linear Accelerator (SLAC) is two miles long and accelerates electrons to about 30 GeV. After it had been running a while a storage ring was added to it. This storage ring was 80 m in diameter and was designed to store electrons moving in one direction around it and positrons moving in the other direction. Both electrons and positrons are stored in bunches. Each bunch holds $\sim 10^{11}$ particles, is a few centimetres long and a millimetre thick. The bunches are arranged to intersect in two straight sections of the device. Because they are moving at about the speed of light they do this several million times per second. This gives a sufficiently large number of electron–positron collisions. Because charged particles radiate energy when moving in a circle, there are four radio frequency accelerating cavities included in the circuit to replace the lost energy and maintain the energy of the two beam constant.

The two straight sections where the beams intersect are used for experiments. In late 1974 Burton Richter and many other scientists were investigating how the cross-section, for the production of hadrons from e^+-e^- collisions, varied with energy and also how it compared with the cross-section for the production of just two muons. At one particular energy, 3.105 ± 0.003 GeV, they discovered an enormously increased production of hadrons – up by more than 1000 times on neighbouring values. Moreover, this resonance was very sharp – its width turned out to be only 67 *keV*! That is only 0.002 per cent of its total energy. Compared with the resonance shown in fig. 2.11 it was a very high and very sharp resonance indeed. Its very sharpness had probably concealed its existence. The sharpness also indicated a comparatively long-lived particle. The resonance width gives ΔE in the Heisenberg Uncertainty relationship $\Delta E \Delta t \gtrsim h$ – so if ΔE is small, Δt, the lifetime, must be comparatively large. And just like the long life of Niu's particle, this long life meant a new quantum number – charm.

Ten days later the SLAC group turned up another such resonance – this time at 3.684 GeV, also fairly narrow, 222 keV, but not as narrow as the first. A little before the joint Brookhaven–SLAC discovery, two Harvard theoreticians had suggested if charm existed, then the 'long' life time it conferred might mean the bound system of a charmed quark and its anti-quark could well behave like a bound system of an electron and a positron (called positronium), or of a positive and a negative muon (called muonium). An electron and a positron behave a bit like a hydrogen atom (which is an electron and a proton). There are two main differences. The electron and positron have equal masses, so one cannot say which is the 'nucleus' of this odd atom. The second difference is more catastrophic: if they get too close to each other, they annihilate! This, of course, always happens eventually. But they last long enough for people to measure their spectra. With muonium there is a similar situation – equal particle masses, possible annihilation, complicated by the radioactivity of both partners. Also, because of their greater mass, their distance apart is much less on average and the energies of their spectral transitions are much higher.

In the charming analogy that Applequist and Politzer suggested, the situation is even more extreme – the masses are even higher, the lifetime before annihilation much shorter. The spectral transitions lead, not to low energy photons in the optical frequency range, but to γ-rays. But they happen. The second resonance the SLAC group found is in fact an excited state of the first and can, and does, emit γ-rays and become the original ψ particle. (The SLAC group called their first particle a ψ meson; the Brookhaven group called it a J meson. It is quite often called J/ψ – but also, quite often, just ψ. Richter and Ting were jointly awarded the Nobel Prize for Physics in 1976.) This charmonium hypothesis – the suggestion that the ψ is a combination of the c quark and its anti-quark $\bar{c}$ – predicted this new spectroscopy beautifully. Both this and the long lifetime of the particles lends credence to the reality of the fourth, charmed quark. In three different colours, of course, and along with its three anti-quarks.

Also it supported the idea of certain particles being combinations of c quarks and other quarks. There would be combinations like $u\bar{c}$, if they were mesons, or udc, if baryons. The particles that Niu and his co-workers observed were presumably of this sort. And so it turned out to be. Many of the possible new combinations have already been identified at the large accelerators. Once again the quark hypothesis was tested and once again found to work, providing, now, that 12 different sorts of quark (and there anti-quarks) exist.

Now this left us with a neat scheme for 'elementary particles'. We had four flavours of quark, u, d, s, and c, each coming in three colours. And, in addition, we had the four leptons. So we had a fine foursquare array like this:

e	u_R	u_G	u_B
v_e	d_R	d_G	d_B
μ	s_R	s_G	s_B
$v\mu$	c_R	c_G	c_B

a kind of mini-mandala, very appealing to any Jungian. Moreover, all these sixteen particles are, as far as we know at present, 'point' particles. That is, if they have an electric charge, it behaves as if located at a point in space, not distributed over some sphere (or other shape). For muons it has been possible to check this down to distances of 10^{-18} m, that is, to about $\frac{1}{1000}$ the radius of a proton. The electron is similarly point-like – so, as far as we can tell, are the quarks. But it is important to note the phrase 'as far as we know at present'. One thousandth of the atomic radius (10^{-10} m) is 10^{-13}. If we had only been able to explore the atom down to that distance we would never have found the nucleus.

However, three years ago we did have this neat 4×4 array of point-like, maybe elementary particles. But, it turned out, nature had not stopped at 16. The accelerator physicists not only found yet another quark; they also found another lepton. This lepton was yet another triumph for the West Coast. Once again the Stanford Linear Accelerator delivered the goods.

Ever since the Bristol group discovered the pion, the muon has seemed like a heavy, unstable electron. The only difference between the two particles lies in their masses and in their having different shades of lepton number. So it seemed possible that they were the lightest two members of a series of leptons – and the SLAC team went looking for the next one up. They knew well that when an energetic electron and positron collided they had a reasonable chance of converting into two muons. They reasoned that if the electrons had enough energy they could also convert into the next member of the family. Tentatively, they christened it tau (τ), for the first letter of the Greek word for 'third'. Also, by analogy with the muon, they expected the tau to be able to decay either one of two ways, $\tau \rightarrow e + \nu_\tau + \bar{\nu}_e$ or $\tau \rightarrow \mu + \nu_\tau + \bar{\nu}_\mu$. So if they produced a pair of tau mesons they would have some cases in which one decayed to an electron and one to a muon (plus various neutrinos). The only thing they could think of that would sometimes produce a similar signature was one of the new charmed mesons, called the D meson. This would also produce a quite different signature on some occasions and so they could calculate the numbers and see if they needed the tau to explain any excess. It turned out that they did, and the tau (after several years work) was discovered. It is a very massive particle – almost twice as massive as the proton (1850 MeV). Its lifetime is 4.9×10^{-13} s. We expect (again from analogy with the muon and the electron) that it will have its own neutrino, ν_τ. So, at this time, the nice 4×4 diagram had suddenly become lopsided.

It did not stay that way for long. Leon Lederman, who is at present director of Fermilabs, and his co-workers had for some time been looking at pairs of muons coming from the collisions of energetic protons. He started with the Brookhaven 30 GeV accelerator and went on to the Fermilab machine, which has now reached 500 GeV energy. He came very close to finding the charmed ψ meson I have already discussed. Having not quite got that he lifted his sights to higher masses and eventually got the yield versus mass curve shown in fig. 4.3. It is the same sort of resonance curve shown in fig. 2.11 that showed the existence of N* baryon, the first known of the very short-lived particles. Also, it is the same sort of curve (though not as spectacular) as that showing the existence of the charmed ψ meson. The curve shows the existence of three particles of rather similar mass. They are, in fact, one particle and two of its excited states – just as was found in the case of the ψ particle. Lederman called the particle the upsilon (Υ) meson. As the ψ is a bound pair of the charmed quark and its anti-quark ($c\bar{c}$), the Υ is a bound pair of another, heavier quark, the b quark (b for beauty, or bottom). The upsilon is very massive, 9.46 GeV, ten times more massive than the proton. Its resonance curve is even narrower than that of the ψ, when measured at the electron–positron machine, DORIS, so its lifetime is somewhat longer than the ψ meson. Its existence shows us that at least three more quarks (b_G, b_R, b_B) exist (and their anti-quarks). It makes it very likely that a further three quarks (t_G, t_R and t_B) also exist, to complete a new 6×4 array. t stands for top (or truth). There is no reason we know for supposing that either the quark or lepton families end with the tau lepton and the top quark.

So the end of all this work with these huge and precise machines is that the quark model of protons, neutrons, pions and the other hadrons is almost certainly correct in its major

Fig. 4.3. The resonance curves found by Leon Lederman and his co-workers which established the existence of the upsilon particles.

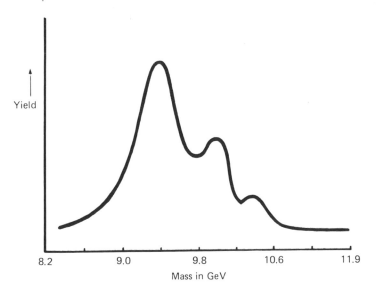

Yield

8.2   9.0   9.8   10.6   11.9

Mass in GeV

ideas. Baryons are made of three quarks; anti-baryons of three anti-quarks; and mesons of a quark and an anti-quark. Always the colours of the quarks add up to white – colourless. But to make the model work we need, *at present*, eighteen quarks (and eighteen anti-quarks). In addition we need the six leptons matching the six quark flavours (u, d, s, c, t and b). And so far, no accelerator experiment has produced convincing evidence that quarks can exist free, outside a hadron.

5

The quest for the free quark

This chapter deals with the search for free, unbound quarks. In the last chapter, and perhaps even more so in this, we have reached the cutting edge of research in particle physics. And as in many similar frontier style activities, things are not always nice, tidy and civilised. One of the main questions we address in this chapter is 'have free quarks been found?'. There are three main answers: 'yes', 'no' and 'maybe'. Of course, many people would prefer their answer to be qualified, such as 'no, I don't think so' or 'maybe, but the evidence is not by any means totally convincing'. But of these three broad categories the answers of most particle physicists today would lie in the final two, 'no' or 'maybe' and the majority of these, I guess, would be in the 'no' class. I, on the other hand, would say 'yes'. The first four chapters, I believe, are fairly unbiassed. I hope this one also turns out that way. But if there are any biasses you now know which way they are likely to lie.

The failure of the early attempts to find free quarks using accelerators at energies around 30 GeV had several consequences. The first of these was to stimulate searches in other areas, particularly in cosmic radiation, where energies much higher than 30 GeV occur. A related area to search is in condensed matter, where one might expect free quarks to accumulate if they came from the cosmic radiation, or were, perhaps, left over from an earlier stage of the Cosmos. Another consequence was to start people wondering if free quarks are 'allowed'. As accelerator energies increased and no certain quarks were found, it began to seem possible that quarks might be permanently confined to the interior of hadrons. In this chapter I shall deal with the experimental searches for free quarks in cosmic radiation and elsewhere, and leave a discussion of their possible confinement to later.

Cosmic radiation (see chapter 1) consists of very energetic charged particles. Its lower energy limit is chosen rather arbitrarily as 10^8 eV. Its upper limit is not known – it is certainly higher than 10^{20} eV. Up to about 10^{15} eV it consists mostly of protons and other heavier nuclei, up to and including iron nuclei. There is a very small component of heavier nuclei still, up to uranium. Beyond 10^{15} eV we do not know for certain what the radiation is: maybe its just the same as below 10^{15} eV; more likely, not. What we do know is that the *energy* is there. The highest acceleration energy is around 10^{14} eV. That is the equivalent energy of the CERN proton–on–anti-proton colliding beam facility. But when the search for free quarks in cosmic radiation started the top machine energy was only 3×10^{10} eV. There is, by cosmic ray standards, quite a large flux of cosmic ray particles with energies higher than that. So in the mid-1960s, many people went looking for quarks in what I would call low-energy cosmic radiation – from about 30 GeV to 3000 GeV.

Most of these searches made several measurements of the ionisation of a particle passing through the 'telescope'. Very often too there was some device that determined the trajectory of the particle. A schematic diagram of one such experiment is shown in fig. 5.1. The experiment was carried out at the University of Tokyo. It had twelve layers of scintillation counter and one streamer chamber. The streamer chamber is a sophisticated device for making the path of the particle visible. It operates like a spark chamber but uses such a fast voltage pulse that the spark only develops to its initial stage of a faint streamer before the voltage is removed. One useful performance figure for these telescopes is called their 'acceptance'. It is the product of the area of the top surface and the solid angle within which the telescope can accept cosmic ray particles. For this device the acceptance is 0.2 m² steradians, which means that the surface area is about 0.4 m² and the solid angle about 0.5 steradians. The solid angle subtended by a sphere at its centre is 4π ($\simeq 12.6$)

steradians. The telescope ran for 3500 hours. *One* track was consistent with that expected for a quark of charge $\frac{1}{3}$e but it occurred before the streamer chamber was brought into action! So the group did not claim to have found a quark. They gave an upper limit on the flux of $\frac{1}{3}$e quarks in the cosmic ray beam at sea level of 0.5×10^{-6} quarks per square metre per second per steradian. The upper limit for the flux of $\frac{2}{3}$e quarks was 15 times higher.

Fig. 5.1. The quark-hunting telescope of Fukushima and his collaborators. This was used to search for single unaccompanied quarks in the cosmic radiation.

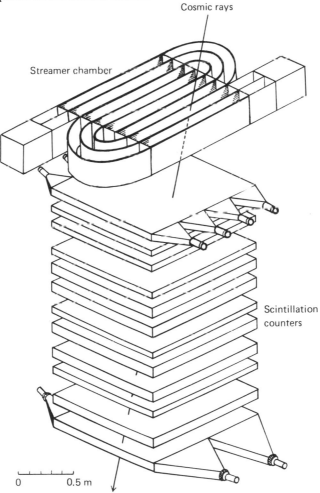

Many other experiments produced rather similar results to this, although this achieved the *lowest* upper limit to the flux of quarks of charge $\frac{1}{3}$e. The general impression was that free quarks had not been found in the low-energy cosmic ray flux. One dissenting result came from New Zealand. P. C. M. Yock and his co-workers reported several instances of fractionally-charged particles using a telescope that measured the time of flight of the particle through the telescope as well as its ionisation. However, my group, using a larger and more instrumented device to repeat the experiment, found no such particles. Combining the results from many experiments, Laurence Jones gave upper limits for the flux of $\frac{1}{3}$e quarks in the low-energy cosmic ray beam of less than or equal to 1.1×10^{-7} quarks per square metre per second per steradian. The limit for the flux of $\frac{2}{3}$e quarks is about twice as high. If free quarks occur in the low-energy cosmic ray beam then there are fewer than given by these rather low limits. So the low-energy cosmic ray results are similar to the results from accelerators. Since the new accelerator results overlap somewhat with these studies using cosmic radiation, this is a satisfactory situation.

My own work in cosmic radiation has mostly been carried out at higher energies, from about 10^{12} eV up to 10^{22} eV (that is, 10^3 GeV to 10^{13} GeV). Some of this work, as I mentioned earlier, was carried out using a large (20 litre) stack of photographic emulsion flown to about 30 kms above Texas and New Mexico. But most of my work has been done with the showers of particles of secondary, tertiary and later generations that are produced by high-energy primary particles when they hit the atmosphere. These showers can cover quite a large area on the ground and so are called extensive air showers (often shortened to EAS).

Figure 5.2 shows the development of an extensive air shower. Up to energies of $\sim 10^{15}$ eV we have some knowledge of the nature of the primary particles from the large emulsion stacks (like my 20 litre stack at Sydney) flown close to the top of the atmosphere. Some of the primary particles reach these stacks without having made a collision with a nucleus in the atmosphere.

We can then measure their ionisation, hence get their electric charge and so determine what sort of a particle they are. For energies between 10^{14} and 10^{15} eV they are a selection of nuclei of atomic numbers from 1 to 26 inclusive (atomic number 1 is hydrogen, and 26 is iron). So we can guess that above 10^{15} eV the composition of the beam is somewhat similar. But we must remember that this is a guess. It is what some people would call

Fig. 5.2. A diagram of a cosmic ray extensive air shower. The original interaction (and many subsequent interactions) produce many pions. If charged, the pions can either interact again or decay to muons. If neutral, the pions usually decay to γ-rays which start the electromagnetic part of the cascade. A single primary particle of $\sim 10^{15}$ eV energy at the top of the atmosphere will result in a shower of between 100 000 and 1 000 000 particles at sea level.

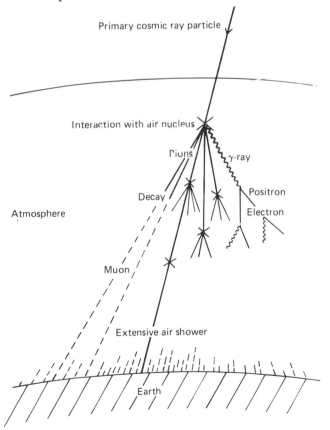

a safe guess – betting that things will not change. But a long time ago Heraclitus said '*Everything changes*'. Anyhow, we can suppose for the moment that the primary particle producing the shower is a proton. When it collides with a nucleus in the atmosphere it will produce a shower of particles, mostly mesons – and most of the mesons will be pions, positive, negative and neutral. Neutral pions have a rather short lifetime, $\sim 10^{-16}$ s in their own rest frame, so most of them decay. Their decay is to two γ-rays. These γ-rays initiate a cascade – each γ-ray produces an electron–positron pair and each electron, when it collides with a nucleus, loses some of its energy and produces a γ-ray. So the process goes on. Very quickly one particle becomes a shower of particles.

The charged pions have a choice. Their lifetime is two hundred million times longer than that of the neutral pion (that is, about 2×10^{-8} s). Moreover, this is their lifetime in their rest frame. In the laboratory (or atmospheric) frame this must be multiplied by their Lorentz factor, their energy divided by their rest mass. If their energy is around 10^{13} eV, this factor is about 700. So their lifetime in the atmosphere is around 10^{-5} s. In this time, because they are moving close to the speed of light they can travel about 4 km. This gives them an appreciable chance of interacting with an air nucleus rather than decaying. So their choice is to interact and spread the cascade, or decay and produce a muon that is so penetrating it will probably reach sea level (and even deep underground). As the mean energy of the pions decreases with each succeeding generation, their chance of decaying to muons increases.

So an extensive air shower consists of three main components. In the centre we have the core of high energy hadrons, maybe including the primary particle. Then around that, densest at the middle and spreading out to about 300 m from the centre, we have the electromagnetic cascade. And interspersed with all of this but spreading out even further are the muons. With our large Pilliga Forest array we have seen muons as far out as 3.4 km from the centre.

As we move from the outskirts of a shower to its centre, the number of particles per unit area (called the density of particles) increases. This is true of all components and is most easily noticeable for the electron component. If the primary particle is a proton then (as can be seen in large emulsion stacks) this density increase continues down to very small distances. Such proton-initiated showers can be seen in these stacks to have this single centre. However, if the primary particle is more complicated, then quite soon after its entry into the atmosphere (maybe on the first collision) it is broken up into its constituent protons and neutrons and each of these starts its own shower. So near the centre of these showers we can have a rather complicated overlapping of the central cores. Roughly, however, there will be as many cores as there were nucleons in the incident nucleus.

As these nucleons move down through the atmosphere they make further collisions with air nuclei. This is part of the spreading of the cascade. But also at each collision each nucleon acquires some sideways (or transverse) momentum. How much it gets depends on the strength of the force that is acting between it and the air nucleus. Because of this acquired transverse momentum the original nucleons gradually move further apart and, because they are the particles that produce the separate shower cores, so do the cores. We can see this process happening right from the first interaction in the large emulsion stacks. Our 20 litre stack is 0.45 m deep. The density of emulsion is close to 4 grams per cubic centimetre so the depth of the stack is 180 g/cm^2, expressed in mass units. That is the equivalent of 18 per cent of the atmosphere, so in the stack we can see a compressed version of an atmospheric cascade. Near the beginning, the separate nucleons are only separated by microns (millionths of a metre) but, as the cascade goes on, the separation increases. When we observe energetic air cascades with a large emulsion chamber (as does the Japanese–Brazilian emulsion collaboration on Mt Chacaltaya in Bolivia (5220 m)) the separation of the cores is measured in centimetres. By the time the shower gets to sea level the separation can be several metres. Just how big this spread is depends on how much transverse momentum the nucleons acquire at each collision. Conversely, if we measure this spread we can get some idea of how big the transverse momentum is – and hence how strong the forces acting are.

When the mean transverse momenta of pions was measured at the early large accelerators (like Berkeley's 5 GeV Bevatron) it turned out to be ~ 0.3 GeV/c (GeV are units of energy: we convert them to momentum by dividing by c, the velocity of light). When a similar measurement was made when 30 GeV accelerators became available, the same result was obtained. Even before that, measurements of transverse momentum had been made in interactions induced by cosmic rays of energy around 1000 GeV – and the same value found. Once the initial surprise wore off, physicists realised that the result was not really surprising. To see why, consider fig. 5.3. In a high-energy collision we generally have one proton sitting at rest in the

Fig. 5.3. This shows the collision of a fast proton and a stationary proton. The fast proton is flattened by the relativistic Fitzgerald–Lorentz contraction.

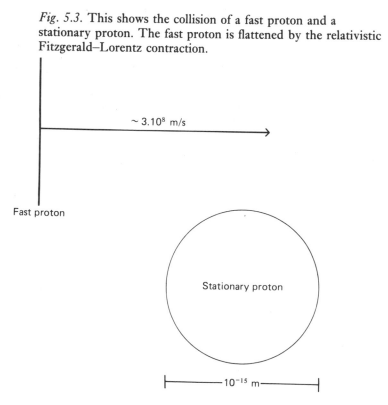

laboratory frame. Its diameter is $\sim 10^{-15}$ m. A high-speed proton approaches it. This proton, because of the relativistic Fitzgerald–Lorentz contraction, is a flat disc (in the laboratory frame). If its energy is ~ 1000 GeV then it is ~ 1000 times flatter than the proton at rest. It is moving at about the speed of light, 3×10^8 m/s. So the two protons are only in contact over the time it takes the disc to pass the sphere, $10^{-15}/3 \times 10^8 \simeq 0.3 \times 10^{-23}$ s. The strong force is a short range force, so that is all the time the strong force has to change the momentum of either proton. And as the energy of the incident proton increases, this time stays almost constant (because the speed is limited to less than the speed of light). So the momentum change, perpendicular to the direction of motion of the proton, remains almost constant. In fact, this transverse momentum can only increase if the force acting increases – because the time of action of the force is effectively fixed.

So it was a great surprise when two Japanese groups (at Osaka City University and Tokyo University) found evidence of much higher transverse momentum at air shower energies (around 10^6 GeV). They did this by the method I have outlined above – measuring the separation of cores in air showers with more than one core. Very soon after, my own group in Sydney confirmed and extended those measurements. For about ten years there seemed to be some scepticism in accelerator circles about these results in cosmic radiation. Then in 1972, the CERN Intersecting Storage Rings came into operation at energies equivalent to 2000 GeV – and there were the high transverse momenta!

Meanwhile, however, I had been looking around to see what could be the cause of this effect. Obviously some force was operating that was much stronger – ten or a hundred times stronger – than the strong force we were used to. At the time there seemed to be two possible candidates. The first was, surprisingly, the weak nuclear force. At normal, 'radioactive', energies this is very weak indeed. But it gets somewhat stronger as energies go up from the MeV region to the GeV region. There was some theoretical possibility that it might get very strong at very high energies. But 10^6 GeV did not seem a high enough energy. The other possibility was the force that bound quarks together to form hadrons: what we might be seeing in the cosmic ray collisions was in fact the result of quark–quark collisions, rather than nucleon–nucleon collisions. So in Sydney we set out to see if either of these two ideas was correct.

By that time we had our large experiment in the Pilliga State Forest working. This was the array of detectors spread out over 60 square kilometres. It was designed to investigate the very high energy cosmic radiation, from about 10^7 GeV up to 10^{13} GeV, and mainly detected muons in the large air showers. It was a fairly easy job to set up some spark chambers and lead in between two close stations. 'Close' meant that they were only 440 yards apart (Australia was still using the old British system of units when we built the array). We then started looking for muons of high transverse momentum in the large showers. Muons are 'weak force' particles and if the weak force was becoming strong it might be expected to show up in the muon component. However, nothing very unexpected happened.

I decided to look for quarks in the core of smaller air showers. The array of close-packed scintillators, which had detected the high transverse momenta, was still operating in Sydney. We had four Wilson cloud chambers left over from other experiments, so we set them up close to the 64 scintillators and gave them the same triggering arrangement. Again, it was a fairly easy thing to do.

Figure 5.4 is a diagram of the layout of the experiment. The 64 scintillators make a chess board-like array, 4 m × 4 m. The cloud chambers were as close to these as we could manage. Three of the four chambers we shielded with lead, believing that quarks were probably penetrating particles. The fourth chamber we left unshielded, in case they were not. The whole array was triggered by the simultaneous discharge of three trays of Geiger counters just under the roof of the hut containing the scintillators and cloud chambers. One only gets such a simultaneous discharge when an air shower hits the apparatus. We chose the size of the

Geiger counter trays to make sure that they discharged when the core of any air shower containing a total of 100 000 particles or more hit the 64-scintillator array. Now, looking at the diagram, you can see that this meant the core might be as much as 5 metres from the cloud chamber. On the other hand, it could be on the cloud chamber. So we expected quite a wide variety of picture types. Figure 5.5 shows four pictures of air showers having varying particle densities in the cloud chamber (these pictures are for unshielded chambers). They illustrate one of the possible difficulties. If event 31(*a*) had contained a quark track, where ionisation was $\frac{1}{9}$ or $\frac{4}{9}$ of the tracks shown, then it would

Fig. 5.4. This shows an overall map of the Sydney 64-scintillator and air shower arrays (*a*) and a more detailed map of the layout of the scintillators and cloud chambers (*b*).

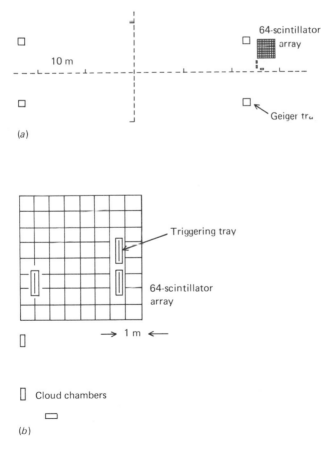

have been easy to spot. If the quark had occurred in 31(*b*) it would have been possible to spot it (but difficult), but for 31(*c*) and (*d*) it would have been quite impossible. As you can see, there are not even any visible gaps between the tracks in 31(*d*).

Fig. 5.5. Four Wilson cloud chamber pictures of extensive air showers. The tracks visible in the first two (*a*) and (*b*) are mostly due to fast electrons passing through the cloud chamber. The third picture (*c*) is of a very dense shower with the tracks very close together. In the fourth picture (*d*) the density is so great that the chamber appears uniformly white. Grossly over-exposing the enlargement however makes some density gradients visible.

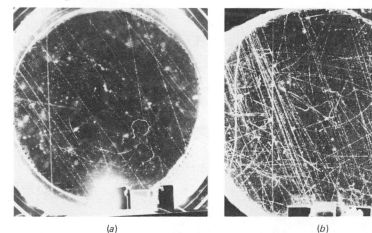

(a)   (b)

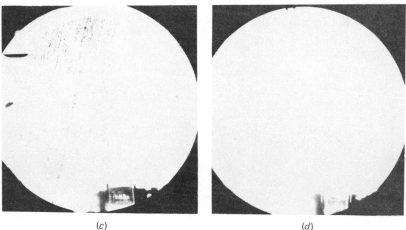

(c)   (d)

However, the cloud chamber was working properly. I tried overprinting the enlargement by 60 times and then I could see some separate tracks at the top of the chamber. 31(c) was produced by the same shower – the cloud chamber was 1 m away from the chamber producing 31(d). About a month after getting these two pictures I remembered that the photographed area of the cloud chambers was $0.3 \times 0.05 = 0.015$ m² and that they had, at that time, been running for about two years. I realised my personal cross-sectional area is a lot more than this and I have been 'running' a lot longer. So quite a number of times in my life I must have received a quite high intensity (but rather brief) burst of ionisation. And that goes for all of us.

To come back to the quarks: fortunately most of the showers produce pictures like 31(a). But in assessing this quark search it is worth remembering that if the quarks occur only in the centre of large air showers they may be difficult to detect.

The four pictures in fig. 5.5 are prompt exposures. That is to say, as soon as possible after the arrival of the shower the cloud chamber was expanded. The delay in initiating the expansion is tiny – for we are dealing with electronic signals. But the actual expansion is a mechanical operation and that takes about 2×10^{-2} s. Once the chamber has been expanded the gas inside is supersaturated with vapour and drops begin to grow on the ions. They were allowed to grow for 10^{-1} s and then photographed. Now if one is interested in measuring fairly accurately the ionisation produced by a particle there is a well-known way of proceeding. Instead of expanding the chamber as soon as possible after the particles have passed through it, one waits for about 10^{-1} s. The ions are originally produced along a very narrow line – only a few atomic diameters wide. Waiting a while lets them diffuse into a much broader column. The vapour then has more chance to condense on separate ions rather than on close groups of ions and one gets a better idea of the ionisation. We tried various delays and finally chose 10^{-1} s as the best for our purpose. We wanted to distinguish clearly between the ionisation produced by a quark and that produced by a normal particle – rather than make an absolute measurement of ionisation. We also needed the chamber to be able to sit inactive for long periods and to behave well when the shower arrived. The showers came in at a mean rate of about one per hour.

In addition to the 64 scintillators, the four cloud chambers and the triggering system, we had four trays of large geiger counters outside the hut, as shown in fig. 5.3. Once we knew the position of the shower centre we could use the response of these trays to enable us to estimate the total number of particles in the shower and hence the energy of the primary particle that produced the shower.

The method of quark detection had two other useful features. The cloud chambers give the directions of the particles through them. As can be seen for fig. 5.5, many of the particles are approximately parallel. Tracks that are not parallel to the main direction are due to low-energy particles scattered in the atmosphere. The quarks in the shower core were expected to have high energy – they might even be fragments of the primary particle. Hence a quark track would most likely be very close to the main direction of the shower.

The other feature that the quark should share with the other shower tracks is its time of arrival. Most of the particles in an air shower move at almost the speed of light. When a shower reaches sea level the great proportion of its particle lie in a disc only two metres thick. They all arrive at the detectors within a few nanoseconds of each other. We then make the tracks wait for 10^{-1} s before expanding the chamber. During this time the ions diffuse outwards and form a cylinder of ions. The radius of this cylinder increases as the square root of the waiting time. This is an example of a law worked out first by Albert Einstein – the problem was called the Drunkard's Walk. We checked it by measuring the widths of tracks obtained with different delays. It worked. So the quark track should not only have the same direction as the other tracks; it should have the same width. We could measure widths from the photographs quite accurately with a microscope fitted with a micrometer eyepiece. A typical value was 1.73 ± 0.02 mm.

To measure the ionisation of the particles we counted the

number of droplets per unit length of track. For normal fast singly-charged particles our chambers gave around 40 droplets per centimetre. The diameter of the chambers was 30 cm, so the tracks were always shorter than this. The ionisation process is a random process so in one centimetre one does not always observe 40 droplets – sometimes it fluctuates up, sometimes down. If the production of each droplet was quite independent of any other, then the standard error of the determination of the ionisation from one track containing N droplets is $N^{\frac{1}{2}}$. This means if we made a large number of determinations of ionisation using tracks of a given length, which gave N drops in that length on average, then 68 per cent of the measurements would give results in the range $N \pm N^{\frac{1}{2}}$, 95.5 per cent would lie in the range $N \pm 2N^{\frac{1}{2}}$, 99.7 per cent would lie in the range $N \pm 3N^{\frac{1}{2}}$. So, for a 20 centimetre long track with a mean ionisation of 40 drops per centimetre only 0.3 per cent of the readings would be either smaller than 35.8 or larger than 44.2 drops per centimetre. Excursions beyond these limits became very improbable quite rapidly.

The situation is somewhat complicated by the fact that the formation of the droplets is not entirely independent. Sometimes the fast particle knocks an electron out of an atom with such speed that the electron itself can knock out more electrons. All these ions are due to the same *primary* collision. The formula for the error has to be adjusted to allow for this. We did it in the obvious (and foolproof) way by just measuring a large number of normal tracks.

So there we had our quark trap. It was a device for detecting fairly energetic air showers and for looking for lightly ionising particles near to their centres. The lightly ionising particle had to be in the same direction as the normal tracks; it had to arrive at the same time; and it had to have an ionisation well outside the normal range of statistical fluctuations. After a bit more experience we added the criterion that there had to be enough normal shower particles in the same event in the cloud chamber to allow a good comparison. We started running in 1968.

After we had been running for a year, one of the local

newspapers found out what we were up to and published a story indicating that we had found the quark. The story was quite widely copied. I was in a quandry. I felt that I was on a promising trail. We had four 'candidates' by that time, but all had defects: too broad, not parallel, unaccompanied – or something. I wrote a paper describing what we had without claiming to have found a quark. Then, while it was still in course of publication, we got the track shown in fig. 5.6. This was a different kettle of fish. It was obviously lightly ionising. It was parallel to many other tracks in the chamber and of the same width. It was well within the illuminated area of the cloud chamber. (All events were photographed by two cameras to give stereoscopic pairs of pictures; these give an excellent perception of depth when viewed in a stereoscope, and depth can be accurately measured with a device called a parallax bar.) The response of the 64 scintillators showed that the particle producing the lightly ionising track had been very close to the core of an air shower containing about 3×10^5 particles. When we measured the ionisation we found it to be only 16.2 ± 2.5 drops per centimetre. The mean for three parallel tracks in the same region of the chamber (which was very close to the overall mean for tracks in many photographs) was 40.7 ± 1.0 drops per centimetre. I wrote a short letter about the event and it was published, very quickly, by the American journal *Physical Review*. Soon after I spoke about it at the International Conference on Cosmic Rays held that year in Budapest – and then was invited to speak at many universities on my way home across Europe and the USA. In Berkeley, California they seemed particularly interested and

Fig. 5.6. A section of the cloud chamber photograph of event 66 240 in the Sydney quark search. The tracks labelled 1, 2 and 3 are due to fast electrons in the air shower. The tracks are wider than those in previous pictures because they were allowed to diffuse before expanding the chamber. This makes it possible to count the number of drops per unit length. The track due to the quark is labelled Q. It is parallel to the other tracks and of the same width but with many fewer drops per unit length. The print has been somewhat over exposed to make the quark track easier to see.

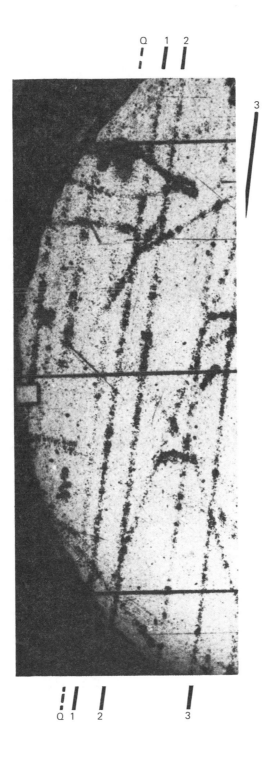

we discussed a repeat experiment they thought they might set up. They had 11 cloud chambers, ready for use but not actually in use, and soon after began to get an experiment going at the Livermore laboratories. Other scientists (at Edinburgh, Leeds and Ann Arbor, Michigan) began cloud chamber searches, and a group at the University of Durham in England began to build a fairly large area detector using a device called a neon hodoscope, rather than cloud chambers.

Meanwhile, a number of people suggested other possible explanations of the track, or gave reasons for believing it not to be due to a quark. One possible explanation is that it is just due to a fluctuation in the number of ions produced by a normal singly-charged particle. We had been able to count the number of droplets over 10.4 centimetres of the lightly ionising track – we found 168 drops. Similar lengths of normal tracks in the same event and the same region of the chamber gave 430 droplets. The chance of observing 168 droplets once, in the course of the whole experiment where 430 are expected, is very small indeed.

However, there is another possibility. A singly-charged particle does not always produce the same number of ions per centimetre. Figure 5.7 shows the way in which the ionisation per centimetre varies with the energy of a particle. The point marked A is called the minimum of the ionisation curve. The flat region at higher energies is called the plateau. However, occasional particles will be at the minimum – they are generally easy to spot because being so low in energy they are noticeably scattered as they move through the gas. Some theoretical physicists have calculated the rise in ionisation from the minimum to the plateau could be as much as 60 per cent. This would mean that a singly-charged particle at this minimum would, on average, produce only 269 drops in 10.4 cm. This is still much greater than the 168 we observed, but a fluctuation from that value is obviously more likely than from 430.

However, from our own experience and from previous experiments, we did not think that a drop of 60 per cent occurred. The group at Livermore settled the matter by making a direct comparison of the ionisation of normal air shower tracks with electrons at the ionisation minimum. They found a drop of only 18 per cent and this means that the probability of our track being due to such a fluctuation is a lot less than a million to one.

So I awaited the results of the various cloud chamber searches with some confidence. But months went by without any other quark track being found. In the end, *none* of the four cloud chamber searches I mentioned found a track that could reasonably be ascribed to a quark.

Two of the experiments (those at Edinburgh and Ann Arbor) were on a small scale and had a much smaller exposure than our own so could easily be explained away. The Leeds experiment used a very large cloud chamber but could not detect quarks within a metre or two of the shower core – which was where we found ours. So that result, too, could be explained. But Livermore used cloud chambers very like ours, just somewhat bigger. Where we used four, they had eleven. Their total exposure was four times as great as ours. It looked very much as if they should have seen a quark, if quarks had been there. For a long time I could see no good explanation for this.

Figs. 5.7. A graph showing how the ionisation produced by a fast charged particle varies with its energy.

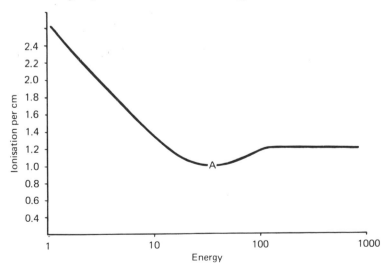

Meanwhile another experiment got under way – the neon hodoscope experiment at Durham under Dr Fred Ashton. As its basic detector this used a very simple device – a long (~ 2 m) glass tube of internal diameter 0.016 m, filled with neon gas. If such a tube is placed between two metal electrodes and subjected to a high voltage pulse just after a charged particle has traversed it and produced some ions in the neon gas, it will glow – just as neon tubes in advertising signs glow. If we make a stack of such tubes and photograph them end on after a charged particle has passed through the stack, then the path of the particle is made visible. For reasonable efficiency, the high voltage pulse must be applied quite quickly after the particle has passed through the stack – within about 10 μs – and this is easy with modern electronics. Such a device is called a 'neon hodoscope' and was first invented by Marcello Conversi in Italy. Figure 5.8 is a diagram of the Durham neon hodoscope. As you can see, it is quite a large device with a much bigger collecting area than our small cloud chambers.

If a singly-charged particle, a muon say, goes through layers F_2 and F_3 of this hodoscope, then it will on average discharge

Fig. 5.8. The neon hodoscope used by the Durham group to search for quarks in the cores of cosmic ray air showers. F_1 to F_8 are banks of neon filled glass tubes ($\sim 11\,000$ in all). The layer of lead above these banks stops the numerous air shower electrons but is penetrated by muons and any quarks present.

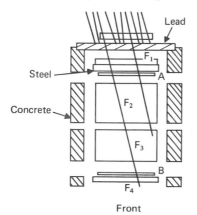
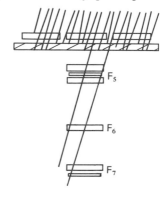

Front

Side

Scale ⌐ 0.5 m

77 of the tubes. However, a quark of charge $\frac{1}{3}$e, because it produces only $\frac{1}{9}$ of the ionisation of the muon, will discharge on average only 46 tubes. This is quite a good distinction between particles, although it is nothing like as good as our drop counting method in a cloud chamber. What Durham gained on the 'area' swings, they lost on the 'distinction' roundabouts. The apparatus contained 11670 tubes altogether! They were shielded by 0.15 m of lead to remove the air shower electrons. So most of the particles traversing the tubes were muons. Now as soon as the ions are produced in neon, they begin to recombine to form neutral atoms. After about 100 μs only about $\frac{1}{9}$ of the original ions are left. So if a chance muon had traversed the hodoscope about 100 μs before the shower came along, it would give a quark-like track in this device (*not* in our cloud chambers). Of course, such a muon would not generally be parallel to the shower tracks, so the Durham group insisted that any quark 'candidate' had to be parallel to the shower tracks before they would consider it. (We had done the same.)

Another aid in distinguishing quark tracks and early, random muon tracks is the number of high energy electrons the particles produced – high enough energy to penetrate the glass walls of two neon tubes and trigger an adjacent tube. Muons, on average produce 1.8 such 'knock-on' electrons in transversing the hodoscope – $\frac{1}{3}$e quarks only produce 0.2. So the Durham group insisted that any quark candidate should produce *no* knock-on electrons.

After running for 15179 hours they had two candidates that satisfied all these requirements. Their total exposure was 15 times greater than ours. There was a 7 per cent chance that their tracks could be due to early random muons – but that explanation does not work for our track. It took a while for me to see the real significance of this result (maybe because I was concerned with another matter I shall deal with later). But then I realised that if I took *all* the experiments, added up all the candidates and then totalled the exposures I would get a value for the flux of quarks in air shower cores. I did this. It came out as 3×10^{-8} quarks per square metre per second per steradian. This flux immediately explained why the Edinburgh, Ann Arbor and Livermore experiments had not seen a quark. With their total exposure they would have been lucky to see one between them. And it was also a high enough flux to explain why we had seen one in Sydney. All we needed was the sort of luck that produces a 10 to 1 win on a horse. And that sort of luck I could believe in.

Meanwhile, other evidence was accumulating that supported a flux of quarks of $\sim 3 \times 10^{-8}$ per square metre per second per steradian in high-energy cosmic radiation. In 1962, a collaboration was started between Brazilian and Japanese scientists. The leader of the Brazilians was Dr Cesare Lattes, one of the co-discoverers of the pion. The two leading Japanese were Dr Yoichi Fujimoto and Dr S. Hasegawa, both of Waseda University, Tokyo. They began to expose large 'emulsion' chambers on Mt. Chacaltaya in Bolivia. These emulsion chambers varied somewhat in pattern but a typical chamber was made of two multi-layer 'sandwiches' of lead and photographic emulsion separated by 0.28 m layer of pitch and wood and a 1.5 m air gap. Figure 5.9 shows a diagram of such a sandwich. The lead-emulsion sandwiches would typically have seven or eight layers of emulsion separated by layers of lead 0.01 m thick. Such an arrangement is very good for detecting high-energy γ-rays. The γ-rays form electron–positron pairs very easily in the lead – the pairs go on to produce more γ-rays and an electromagnetic cascade builds up rapidly. When the next layer of emulsion is reached, the large number of positive and negative electrons sensitise the emulsion. When it is developed they leave a little black dot. The dot is big enough to be seen by the naked eye if the γ-ray energy is greater than about 500 GeV. An individual cascade can be followed from layer to layer (which gives the direction of the initiating γ-ray accurately) and its energy determined by the way the size of the dot changes.

The pitch is useful for detecting any energetic hadrons that accompany the γ-rays. These hadrons can make nuclear interactions in the lead and emulsion sandwich but they have more chance of doing so in the pitch. These interactions include

neutral pions in their products. The neutral pions decay to two
γ-rays with a short half life ($\sim 10^{-16}$ s) and these γ-rays can be
picked up in the lower lead–emulsion sandwich. These chambers
are very simple and very elegant devices. They need, however,
a lot of lead, a lot of emulsion, a lot of high altitude stamina,
a great deal of emulsion processing, much dedication and almost
infinite patience.

In 1971 the collaboration reported a very unusual event,
which they called Centauro. It is shown diagrammatically in fig.
5.10. It was first picked up as an event in the lower sandwich
layer. Here they found many energetic cascades due to γ-rays
of total visible energy 220000 GeV. Another way of writing this
is 220 TeV, where 1 TeV = 1000 GeV. They looked in the
upper chamber for the corresponding large showers and for a
while could find *nothing*! Eventually they found an almost

Fig. 5.9. A cross-section through part of an emulsion chamber of
the sort used by the Brazilian–Japanese emulsion collaboration.

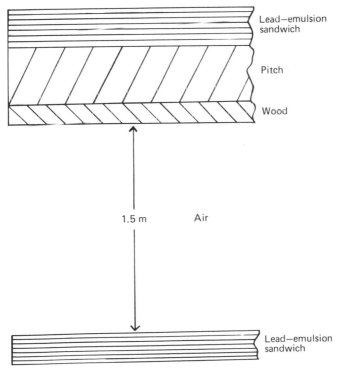

negligible event with a total visible energy of only 27 TeV. Hence the name Centauro – the appearance of the back half had given them no clue as to the appearance of the front half.

This discrepancy between the large numbers of energetic γ-rays in the lower chamber and very few in the upper chamber meant that in the interval between the two sandwiches there had been a large number of nuclear interactions – mostly in the pitch and wood. This in turn meant that there had been a considerable number of energetic hadrons hitting the top of the chamber – and

Fig. 5.10. A schematic diagram of the event 'Centauro'.

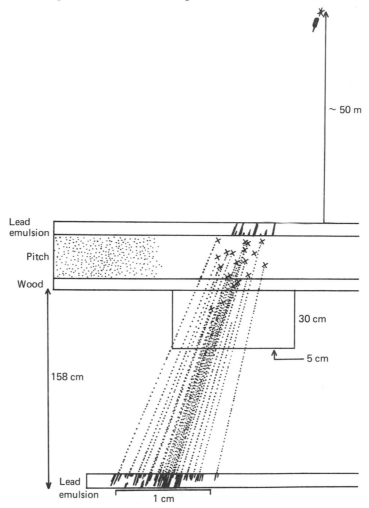

almost no γ-rays. The Japanese and Brazilians were able to trace some cascades back from the bottom layer to the top and they found that they were converging back to a point in the air about 50 m above the chamber. From the number of nuclear interactions they found that a beam of about 75 high-energy hadrons had originated at this point and that *none* of the hadrons were neutral pions (which decay quickly to γ-rays). In normal high-energy interactions many pions are produced (they form the great majority of produced particles). On average the pions are one third positively-charged, one third negatively-charged, and one third neutral. This 'charge independence' had been checked in very many experiments and was built into all theories, including the quark model. So the hadrons in the beam could not be pions. But no one had seen a large, high-energy interaction that did not produce pions and no theory predicted such a thing. Centauro was something new and quite unusual.

Many people (including me) tried to explain it. At one time (when I was working in Tokyo) I thought the explanation was the collision of a very energetic iron nucleus with an air nucleus just 50 m above the emulsion chamber. I went into Professor Miyake's room and wrote on his blackboard '*Centauro is an Iron Horse*'. But it turned out that I had made a simple error in working out the probability of an iron nucleus penetrating that far into the atmosphere and I had to retract the statement.

Then, in 1978, James Bjorken and L. D. McClerran suggested that the 'particle' that caused the Centauro event was a 'quark glob'. They looked at the event with fresh eyes and saw that what was needed was something that could penetrate one half of the atmosphere (equal to 0.5 m of lead) and *then* produce 75 baryons (and no pions). The hypothetical quark glob could do just this. If its size was right it could have enough quarks to produce 75 baryons (about 225 quarks are needed), and if it had only quarks in it (and no anti-quarks) it could not produce pions. Moreover, because quarks bind together very strongly, it would be super dense, hence much smaller in cross-section than a normal nucleus and better able to penetrate the atmosphere. Bjorken and McClerran postulated that these 'quark globs' form a part of the incident cosmic radiation at energies around 10^6 GeV ($\equiv$ 1000 TeV). Maybe they are remnants left over from an early stage of the Universe when theoreticians believe that all matter was in the form of a superdense 'quark soup'.

Now baryons are made of three quarks. If the number of quarks in the glob is an integral multiple of three then, when the glob breaks up, there will be no quarks left over. But when the number of quarks in the glob is not an integral multiple of three there will be one or two quarks left over. These quarks have the same high speed as the glob so they will tend to stay in the centre of the air shower that develops from the baryons produced in the original break-up. So maybe the quark-like tracks that we and the Durham group had seen were made by the left over quarks after Centauro interactions of quark globs. I worked out the flux of events from the Japanese and Brazilian results and found that it agreed with the flux from the cloud chamber and neon hodoscope experiments.

Various other unexplained effects in cosmic radiation also fitted nicely into the 'quark glob' picture. The P. N. Lebedev Physical Institute in Moscow run a large cosmic ray research installation in the Tien Shan mountains. They had been investigating the properties of the core regions of air showers in the energy range above 10^5 GeV. They found, in some showers, particles whose mean free path between interactions was enormously long, about three times longer than that for energetic protons. This is the expected value for quarks on the naive quark model. The argument is straightforward: a proton is made of three quarks so should have roughly one third of the interaction mean free path of a single quark. Again, when I worked out the flux of the Lebedev events it agreed with the flux of quarks from the cloud chamber and neon hodoscope experiments.

So the evidence for the existence of free quarks in high-energy cosmic radiation piled up. Meanwhile, quite a different sort of experiment had produced some very interesting results.

This experiment was one of the searches for quarks in normal matter. Many experimenters had looked without success and many of these had used a variation of Millikan's method of determining the charge of a single electron. Millikan had carried out his experiments around 1910 when the idea of a smallest-possible electric charge was around but had not been established. He had the idea of making very small drops of oil by blowing oil through an 'atomiser' into the space between two horizontal metal plates, which could be electrically charged to some known voltage. He then ionised the drops by briefly illuminating them with X-rays. He could then balance the downward pull of gravity on a drop with an upward pull due to the electric field acting on the charge on the drop. Once he had measured the mass of the drop by determining its radius, he could calculate its charge. He found that almost all of his measurements of charge were integral multiples of a certain very small charge – and this he gave as the charge on a single electron. Note the 'almost' in that sentence. Recall that in his paper of 1910 in the *Philosophical Magazine* Millikan wrote

> I have discarded one uncertain and unduplicated obser-
> vation, apparently upon a single charged drop, which
> gave a value of the charge on the drop some 30 per cent
> lower than e (the electronic charge).

So one of Millikan's observations was of a charge close to $\frac{2}{3}e$ – the expected charge on u quarks! Millikan was awarded a Nobel Prize for Physics in 1923 for his determination of e. It is possible that, in the same work, he also made the first observation of a free quark! However, Millikan's experiment has been repeated many times – it is a standard experiment in physics teaching laboratories. Many values for e have been obtained but I have not seen another value of $\frac{2}{3}e$, using the Millikan oil drop technique, in the literature.

Millikan's oil drops were very small. Their mass was around 10^{-14} kg. When people decided to look for quarks in this way they felt that they would have more chance of finding them if they used bigger test objects. So various ways of levitating larger spheres were considered. A group at Stanford University in California under Professor William Fairbank, decided to do a Millikan-stype experiment using superconducting levitation.

Probably one reason for their choice was that they were very expert at low-temperature experiments using superconductors. This decision meant they had to pick a material that became superconducting at a reasonable temperature. 'Reasonable' here means around 4 K – just four degrees above absolute zero!

When a metal goes superconducting it loses all electrical resistance. When an electrical current is started in such a conductor it will (if conditions stay the same) continue for ever. The current of course produces a magnetic field and that too persists indefinitely. It is possible to arrange things so that a small sphere of superconducting material is magnetised in this way and is levitated permanently in a suitably-shaped magnetic field – also produced by a superconductor. The whole array can be set up in a vacuum, so one has near ideal conditions for measuring an electric charge on the sphere. The Stanford group chose to do this by applying an alternating potential of 3000 volts to parallel horizontal metal plates above and below the levitated sphere. If the frequency of the alternating potential is adjusted to a suitable value it starts the small sphere oscillating. The amplitude of the oscillations can be measured very accurately by a device called a SQUID – a Superconducting Quantum Interferometric Device. Then when the sphere is oscillating, the phase of the alternating potential is reversed, so that it is pushing on the sphere when formerly it had been pulling. The rate at which the amplitude of the oscillation decreases is measured. This rate depends on the force acting on the sphere and this in turn depends on the electric charge on the sphere. If everything else is known, the charge can be determined.

Figure 5.11 is a schematic diagram of this device. When the small sphere is levitated it usually has a small electric charge on it (a few thousand times the charge on the electron). This can be reduced by bombarding it with electrons or positrons from two radioactive sources, which can be swung close to it. If it does not contain a quark this charge can be reduced to zero. However, if there are quarks in the sphere (and their number is not an integral multiple of three) then it will not be possible to reduce the sphere's charge to zero; its smallest value will be $\pm\frac{1}{3}e$.

Hebard and Fairbank first reported results on their niobium spheres in 1970. They had found fractional charges of $\pm\frac{1}{3}$e on some spheres. The results were criticised. They repeated the experiment, with improvements, and got the same result. This cycle has been repeated several times. Figure 5.12 shows the results of their charge measurements in historical order, up to 1981. The occurrence of fractional charges of $\pm\frac{1}{3}$e is obvious. One obvious feature of the results is also clear from the figure. Ball 6, for instance, has at different times had residual charges of zero and $\pm\frac{1}{3}$e. It behaves as though it was losing its quark (or quarks) from time to time.

Professor Luis Alvarez (who was awarded the Nobel Physics Prize in 1968) has made an interesting suggestion for a further test of the reality of these fractional charges. So far, the data for a run has been fed into a computer, which then prints out the various conditions and the results of the calculation including the value of the charge. If anything had obviously gone wrong during the run, it was discarded. But skeptical physicists always

Fig. 5.11. A schematic diagram of the Stanford quark search experiment. A and A^1 are circular, optically flat plates to which an oscillating electric potential can be applied. The test object is N, a small niobium sphere which can be levitated in the field of the superconducting magnet coils M. S is a SQUID, a superconducting device that measures the position of N with great accuracy. The whole apparatus is inside a cold enclosure at a temperature of 4 K – 4 degrees above absolute zero. Not shown is a device for loading different niobium spheres into the measuring position and radioactive sources of positrons and negatrons, which can be swung in to reduce the residual charge on the sphere to zero or $\pm\frac{1}{3}$ e.

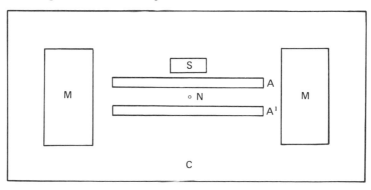

wonder about the possibility that the experimenter might unconsciously be influenced to decide that 'something had gone wrong' if the charge turned out to have some 'undesired value' such as 0.50. In order to avoid that possibility, Professor Alvarez suggested that the computer programme be altered so that a particular random number between 0 and 1 be added to the crude data at a point in the calculation where it would only affect the final value of the charge. Also, that the programme be so constructed that if anyone attempted to 'uncover' the random number the programme would, so to speak, blow up – unless they used a code word that only Professor Alvarez, or some other

Fig. 5.12. The results of the Stanford experiment measuring the residual electric charges on small niobium spheres. ± 0.333 ($= \frac{1}{3}$) is the expected measured charge due to a quark. The results are given in historical order, starting at the bottom. The number of the sphere measured is given at the left-hand side. It can be seen that some spheres were measured several times – not always with the same result.

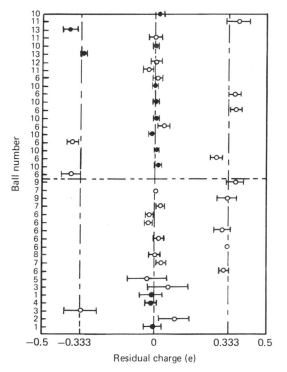

'non-involved person' knew. Once the print out was available and a decision had been made as to whether or not the run was acceptable, that point would be plotted at its indicated (but wrong) charge. When the resulting graph had accumulated perhaps 25 such 'charges', it would be distributed to the physics community, in a pre-print. It would, of course, show none of the 'blank regions', between $-\frac{1}{3}$, 0 and $+\frac{1}{3}$, that the typical Stanford graphs show because when each run was completed, the computer would generate a new random number to be added to each measurement in the next 'run'.

Then, according to the Alvarez scenario, he would be invited by Fairbank to furnish the key to the computer, so that it would type out a list that would say, for example, 'in run 1, add 0.125; in run 2, subtract 0.336....' That list would then be distributed to all those who received the 'scrambled' plot, and each recipient could then use a millimeter scale to move the 25 points to the right and left, by the proper amounts. If the 'unscrambled' plot then showed the familiar 'blank spaces', Alvarez feels that the Fairbank results would then gain wide acceptance. He further says, and Fairbank agrees, that if the experiment is really detecting quarks the random number technique should show even more pronounced blank spaces than previously published ones do. The reason for this is that Fairbank is under great pressure to include *all* his measured charges in his plots. (The first question he is always asked after a talk on his results is, 'how many points did you leave out?') But in this 'blind protocol', Fairbank could throw out any run in which, for example, the 'patch effect curves' changed with time. No one could criticise him for throwing out *any* points, because he would not know at the time what the calculated charge actually was. The consensus is that, if quarks are really being observed on the niobium balls, they should show up even more clearly when the blind technique is being used. This sort of test is often used in medicine in testing drugs on human patients. It is called a 'double blind' test. When I visited Stanford in August 1982 they were setting up their apparatus to carry out a series of runs in this way.

Meanwhile, other workers had been examining other substances. Most of them found no evidence for fractional charges. In the case of iron, the results have been contradictory. An American group found fractional charges; groups in Russia and Italy did not.

After I had noticed that the fluxes from different cosmic ray experiments were similar I decided to see what was the concentration of quarks in the Earth that this flux entailed. The high-energy quarks will generally penetrate some tens of metres into the Earth's crust before they are stopped. If they are heavy quarks they can decay to lighter quarks – but in the end, because of the conservation of electrical charge and their own fractional charge, we must be left with the lightest quark. There is nothing it can decay to and still conserve charge. If it is positively charged it will wander around – repelled by all nuclei (because they are positive) and unable to form a neutral atom (because of its fractional charge). If it meets its negative anti-quark it can annihilate, but as we will see immediately, the negative anti-quarks are not available. The negative quarks are *attracted* to the positive nuclei. Because they are massive they take up orbits *inside* the nuclei. They reduce the nuclear charge by $\frac{1}{3}$e or $\frac{2}{3}$e; but overall it is still positively charged and so repels any positive anti-quark. So both sorts of quark are in a very safe situation and will persist for very long times.

Now the age of the Earth is about 4.5×10^9 years. Geologists have a fairly good idea of what is called the 'mixing depth', the depth from the surface within which rocks are pretty well mixed by geological processes – plate movements, erosion, and so on. Assuming that the flux of cosmic radiation has remained more or less constant over that time, it is easy to work out the expected number of quarks per gram of matter around us. It comes out as 0.036 quarks/gm. The Stanford group need at least ~ 1 quark per ball to explain their results. The mass of the balls is $\sim 10^{-4}$ gm so that the expected density of quarks from their results is $\geqslant 10^4$ quarks per gram. This is a much higher concentration than the cosmic ray flux implies.

There are a number of possible explanations of this. One is

that the Stanford quarks do not come from cosmic radiation but are themselves a minor contaminant of ordinary matter, left over from the Big Bang. Another is that somehow quarks are concentrated in niobium by geochemical processes. Geochemical processes can produce some remarkable concentrations. The concentration of uranium in a rich uranium mine is about 100000 times greater than the concentration of uranium in the Earth's crust as a whole. That factor is just about the ratio between the Stanford concentration and my calculated concentration for the quarked cosmic ray flux so, it is possible. At present there are quite a number of ingenious Millikan-style quark searches in progress looking at a variety of materials.

To summarise these last two chapters: almost all physicists in the field believe not only in the existence of quarks as constituents of hadrons but believe that 15 types of quark are already known and a further three types are almost certain to exist. But many physicists at present think that quarks are confined to hadrons; some physicists are quite dogmatic about this, others are prepared to suspend final judgement but consider it to be likely. A minority consider it likely that free quarks exist, and some (including myself) believe the evidence for their existence is very good. The matter is obviously open to experimental test. A flux of around 3×10^{-8} quarks per square metre per second per steradian in the cosmic radiation means that one would get about 20 quarks per year on a sensitive area of 20 m². That is not a large area by the standards of present day accelerator experiments and already some experiments are under way.

6

Matter and the Universe

In this chapter I shall tie up a few loose ends and then have a look at the broader implications of all the detailed and complicated work I have described.

First I shall deal with quark confinement. The enormous accelerators at CERN and Fermilabs have not produced quarks in any considerable numbers; in fact most people doubt if any have been produced. So quarks are at least very difficult to extract from nucleons. They are bound (or confined) very strongly in them. Some eminent theoreticians think they are absolutely confined, others are not so sure. But, at least, they are confined to a very considerable extent. Yet when we examine the quarks *inside* a proton or neutron they appear to be moving around freely – 'asymptotic freedom', it is called. How can this be? The answer turned out to be easy, but surprising.

The first force field physicists learned about was gravity. Now the force between two gravitating objects decreases as we move them apart. It is a rapid decrease, getting smaller as the square of the distance between them. If we double the distance, the force is less by a factor of four. When the electrical force was investigated it was found to obey a similar law – the force between two electrical charges varies inversely as the square of the distance between them. There is a difference between the two fields: gravity is always attractive, while the electrical force can be either attractive or repulsive, depending on whether we are dealing with unlike or like charges. But whether it is attractive or repulsive it is always 'inverse square'.

When Yukawa put up his theory of the strong nuclear force he took over the inverse square law, adding an extra multiplicative term. This term varies with the separation as a negative exponential; this means that outside some characteristic distance (which Yukawa took to be $\sim 10^{-15}$ m) the force decreases very quickly.

All these forces have one thing in common: they decrease as the separation of the particles increased. But this is not true for *all* forces. For instance, if we fasten two spheres together with a length of elastic we have a situation where the force between the spheres is zero as long as the elastic is slack and then increases with the separation of the spheres. And it keeps on increasing until the elastic snaps.

It looks as though the force between quarks is similar. This force produces 'confinement' and 'asymptotic freedom'. Whether it produces absolute confinement depends on whether the quark force is like the elastic (which snaps eventually) or whether it does not.

Any quantum theory of forces needs some particle to 'carry' the force. For gravity, they are gravitons; for electromagnetism, photons; for the strong force, pions and other mesons; and for the weak force, the intermediate vector bosons. The quantum theory of quark forces is called quantum chromodynamics (or QCD, for short). It operates with 'colour charges' to match the colour of quarks (hence the 'chromo' in chromodynamics). It needs *eight* carrier particles, called gluons. It appears that if the gluons are like photons and have a zero rest mass, then the range of the force is infinite (the 'elastic' never snaps). Otherwise its range is not infinite, and confinement is not absolute. So far, it has not been possible to prove, using QCD, whether confinement is absolute or not.

But notice now that we not only need at least eighteen quarks but also eight gluons to explain the phenomena of physics. And also six (at least) leptons, plus the photon, the graviton and the intermediate vector bosons. Once again, our list of 'fundamental building blocks' is becoming uncomfortably large. It looks as though the rainbow has tricked us again and the crock of gold is elsewhere.

This multiplication of quarks has led to various theoretical physicists proposing schemes of sub-quarks, hoping that with them we may reach the few really simple and fundamental building blocks of nature. But others are beginning to wonder if this is a sensible approach. They remember that we have seen quite a number of supposedly 'fundamental building blocks' that proved not to be so. The allegedly indivisible atom turned out to be very divisible, into electron cloud and nucleus. The nucleus itself was found to be structured, made of protons and neutrons held together by pions and many other mesons. The protons, neutrons and pions are made of quarks and anti-quarks. One begins to wonder if the question 'what are the fundamental building blocks of the Universe?' is a sensible question. It is only sensible if one *believes* that the Universe is basically atomic, that it is made up of a very large number of a few species of indivisible particles. It may be worthwhile to look at the possibility that the Universe is *not* basically atomic.

Straightaway we find that until quite recently (say about AD 1870) belief in the atomic theory has been very much a minority opinion. The great majority of the Greek philosophers were against it – some of them violently so. Plato wanted the books of Democritus burned. It is not a method of philosophical discussion I recommend but it does show the strength of his feeling. The Middle Ages hardly bothered even to consider atomism. With the rise of modern science the belief became somewhat more popular, but remained a minority view. Newton was, in a sense, an atomist. He believed it likely that the material Universe is made of atoms. But for Newton the material Universe is a minor manifestation of Spirit. He devoted a great deal of his time to research in alchemy and theology. I suspect that this alchemical work was in the great tradition of alchemy. Carl Gustav Jung and his school have established the true significance of this tradition in European thought. Newton's picture of the Universe was certainly not 'atoms and nothing else'. In fact he protested strongly against the 'notion of bodies having, as it were, a complete, absolute and independent reality in themselves'.

Outside of Europe, the great philosophical schools of India and China hardly considered atomism. It is not that the possibility escaped their attention; rather, they saw at once that it would not work.

So up to about 1870, the atomic hypothesis was not highly favoured. However around that time the application of Dalton's and Avogadro's ideas led not only to considerable progress in chemistry but also to the kinetic theory of gases in physics. From then until 1925 the atomic idea flourished. But in 1925 quantum mechanics came in. The scientists who began quantum mechanics, Schrödinger, Heisenberg, Bohr, Pauli and others, had all been brought up in the atomic tradition. They found it very hard to believe the Universe is as the experiments say it is. Heisenberg, in particular, has graphically described how crazy the new theory seemed to him. But it worked – and no other theory did. One might say the facts pushed the scientists into quantum mechanics.

Quite soon after the discovery of quantum mechanics, Heisenberg noted that it is anti-atomic. The quantum-mechanical model sees the Universe as one, whole and indivisible, not something made up of a very large number of small and indivisible entities. We saw in chapter 1 that this is not the only unusual feature in quantum mechanics. The observer (namely, *us*) forms an essential part of the quantum-mechanical description of the Universe. Remember Schrödinger's cat and the conclusion, that in the quantum-mechanical picture things do not happen between observations. Remember too the Uncertainty Principle, graphically demonstrated by our complete inability to say which of two muons, produced in the same interaction, will decay first, or even when either of them will decay. These features of quantum mechanics were very hard to swallow. It was only that, in spite of them (or maybe because of them) quantum mechanics makes astonishingly accurate predictions (always in the form of probabilities, of course).

Even so, the theory was too much for Einstein to swallow. In 1935 he published a paper (along with Boris Podolsky and Nathan Rosen) in which he thought he showed that, although quantum mechanics was a good theory, it was not a complete theory. He believed that the motions of particles were uncertain not because they were in reality uncertain, but because some parameters that determine the motion were not yet known to us – that there were some 'hidden variables' in the problem.

Einstein reached this conclusion by using one of the 'thought experiments' for which he is famous. One version of this experiment (given later by David Bohm) goes as follows.

We imagine some atomic or nuclear state that has zero angular momentum (called spin for short). This state decays to two similar particles, each with a non-zero spin. Because angular momentum is conserved, the total angular momentum of the final state must be zero. So if one of the two particles that make up the final state has spin $\frac{1}{2}$, the other must have spin $-\frac{1}{2}$. We can imagine this state of affairs by looking at fig. 6.1. Now we can measure spin in quantum mechanics around any one of the three mutually perpendicular axes, x, y or z. But we cannot measure it simultaneously about more than one axis. Also, one of the peculiarities of quantum mechanics is that no matter which way we point the axis we are going to use, we shall only ever get $\frac{1}{2}$ or $-\frac{1}{2}$ for the answer.

Suppose we measure the spin of particle 1 about the x-axis and get the value $-\frac{1}{2}$. Then we know without making a measurement on particle 2 that its x-component of spin must be $\frac{1}{2}$. Because its x-component is $\frac{1}{2}$ and we have not disturbed it, its x-component must have been $\frac{1}{2}$ before the measurement on particle 1. Now we could just as easily have measured the y or z components of spin of particle 1 and have found the spin components of particle 2 without disturbing it. So particle 2

Fig. 6.1. The Einstein–Podolski–Rosen experiment envisages an original state consisting of a single particle with zero spin (*a*). This decays to two spinning particles (*b*) whose total spin must add to zero by the law of conservation of angular momentum.

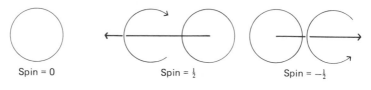

Spin = 0   Spin = $\frac{1}{2}$ Spin = $-\frac{1}{2}$

(*a*)   (*b*)

must have had a definite spin before any measurements were made on particle 1. But this is certainly not included in the quantum-mechanical wave function of the system. Therefore quantum mechanics is an incomplete theory.

This was a very serious criticism of quantum mechanics and seemed for a while to be at least very difficult to answer. Then in 1965 John S. Bell of CERN published a short paper on the experiment. What Bell did was to use the fundamental assumptions that Einstein had used to predict counting rates in certain feasible experiments in atomic and nuclear physics. He noticed that for some of these experiments quantum mechanics predicted different rates to the 'Einstein' rates. This set the stage for a standard procedure in science, an experimental test of two rival theories.

It has proved possible to do the experiments in a variety of ways. One way used low energy photons of visible light; another used high energy γ-ray photons from the annihilation of positrons and negative electrons; a third used *protons* of medium energy (around 10 meV). The experimental results have been decisively in favour of quantum mechanics and in contradiction with the Einsteinian prediction.

So, because Einstein's logic was impeccable, we must look to his basic assumptions. One at least of these must be wrong. These three assumptions are

(1) Observed phenomena are caused by some physical reality whose existence is independent of observation. Sometimes this is called the assumption of 'realism'. Roughly it means that if you are alone in a room with a table and shut your eyes, the table does not disappear (but who is to know if it did?).

(2) Often called the assumption of induction, this says that inductive inference is a valid mode of reasoning and can be used to draw legitimate conclusions from consistent observations. It has been impossible to prove logically that this is so, but most of science is based on it.

(3) This is called separability or locality and says that action at a distance does not occur and that no influences propagate faster than light.

If we drop any of these assumptions we destroy the 19th-century 'scientific' materialist view of the Universe. If we cannot use induction universally, then most of the scientific 'discoveries' cannot be taken as proven. If we allow influences that move faster than light, causal chains break down because what is 'cause' in one frame of reference becomes 'effect' in another. And if the first assumption is not correct we are dealing, not with a realist Universe, but with a Universe in which the observer is essential.

Note that these experiments not only showed that at least one of Einstein's assumptions was wrong; they showed also that the prediction of quantum mechanics were right. This, of course, is not always the case in such tests. Sometimes (to the delight of some experimentalists) experiment shows that *both* theories are wrong, 'a plague on both your houses'. But in this case it was not so and once again quantum mechanics came through the test with flying colours. In effect, the Einstein–Podolsky–Rosen experiment has been carried out and has given the opposite answer to what they expected. Quantum mechanics stands as the best description we can give, so far, of the physical Universe.

Let us look at this theory and what its originators thought of it. First, it is anti-atomic. It pictures the Universe as one, whole and indivisible. This is not to say that the idea of chemical 'atoms', for instance, is completely false. Such entities as the hydrogen and oxygen 'atoms' exist. But they are *not* fundamental building blocks. The are more like eddies on the surface of a river. The Universe is seamless, but not featureless. Secondly, if the Universe is one, whole and indivisible, then because you and I are part of it, it is one, whole and indivisible, and *conscious*.

These considerations were apparent to the founders of quantum mechanics. They could not help noticing the similarity between these conclusions from modern physics and the prior conclusions from the great Eastern philosophies, Indian, Chinese and Japanese (or, indeed, of any mystical philosophy). Schrödinger said that what physics needed was a good dollop of Eastern mysticism. Heisenberg noted the relationship between the two; Bohr said that modern physicists were confronting problems already dealt with by thinkers such as the Buddha and Lao Tzu (he put the Chinese Yin-Yang symbol in his coat of arms when he was knighted); Pauli got together with Carl Gustav Jung to write a book on *The Interpretation of Nature and the Psyche*.

If we are going to put the observer into physics, obviously it is common sense to study the observer. Serendipitously, other western scientists began to do this at just the same time that physicists were getting into quantum theory and relativity. In 1895 Röntgen discovered X-rays; in 1900 Planck put forward his quantum hypothesis; in 1905 Einstein published his paper on the Special Theory of Relativity. In the middle of all this, in 1900, Sigmund Freud published *The Interpretation of Dreams*. The two sciences of physics and psychology at that time seemed very far apart. But in an indivisible Universe this distance is more apparent than real. Since that time Western psychology (like Western physics) has advanced very quickly (it had a lot of Eastern psychology already there to draw on). It has gone from Freudian, to Jungian, to humanistic to transpersonal. (These are very broad categories – there are many sidetracks and subdivisions.) Now, in the study of consciousness, it has begun to incorporate physics!

At one time it was quite popular to try to explain consciousness as a by-product of biochemical interactions in the brain. Stanley Jaki, in his book *Brain, Mind and Computers*, points out that this attempt was not popular with the leaders in the field (his examples include Sir Charles Sherrington, Sir John Eccles and John von Neumann). But the epigoni (a word I had to look up) liked it. Developments in both fundamental physics and modern psychology seem to me to support the opinions of the masters rather than of the epigoni.

So let us have a brief look at some of the findings of modern research with consciousness. This was certainly well outside my field of expertise eight years ago. In the last eight years I have done a fair amount of reading in the field and a reasonable amount of experimental work (using myself as the experimental animal, of course). But my training was as a physicist and so here I rely mostly on the work of others. Ken Wilber has summarised and synthesised this work admirably in three books (and a number of articles). The books are *Spectrum of Consciousness* (1977), *The Atman Project* (1980) and *Up from Eden* (1981). These books list five or six hundred references for further reading.

One of the ideas I found a little hard to accept at first is that consciousness comes first – that in fact everything is consciousness. This is very much a reversal of the Cartesian/materialistic idea that consciousness is somehow a by-product of the biochemical activity of the material brain. It says, in effect, that matter is the lowest level of consciousness.

A little thought convinced me that this could well be true. I started by noting that I am conscious – that I can feel, have emotions, think, and so on. Then I noted that animals, dogs for instance, also appear to be conscious. A dog certainly seems to feel love, anger, joy, shame and guilt (as in a hangdog look), lust and so on. But it does not have the kind of reflexive consciousness that I have. It does not say 'I am a conscious dog'. So it is conscious, but not at as high a level of consciousness as I. Other animals also seem conscious. A lizard, for instance. But the lizard does not show the range of emotions that a dog does, so it is at an ever lower level of consciousness. An earthworm is at a yet lower level, though still obviously conscious. It responds quite actively to a variety of 'outside' stimuli. Some plants, the Jamaican sensitive plant or the Venus fly trap for instance, respond quite rapidly to some stimuli (though not as many as the earthworm). They also grow, reproduce and so on. They are obviously alive, but again at a still lower level of consciousness. One can go from these in stages down to single cells. There is never any dividing line that we can say separates consciousness from non-consciousness. And from cells to virus particles, which are large molecules, and so to smaller molecules – to matter in fact, as the lowest level of consciousness.

Wilber distinguishes eight main levels of consciousness. These are Nature (the lowest level, including matter and lower life forms); Body (higher bodily life forms); Early Mind (man up $\sim$ 1500 BC); Advanced Mind (rational, mental–egoic, the state most of us are in now); Psychic; Subtle; Causal; and Ultimate. Within each of these there are many sub-levels. For instance, it is generally supposed by physical cosmologists today that at a very early stage of the Universe, soon after the Big Bang, all matter was in the form of 'quark soup'. This would be about the lowest form of consciousness on the Wilber scheme. It is also interesting to notice that modern physics was 'pushed' into the Big Bang theory of the evolution of the Universe by facts (though not all astronomers are as yet convinced). But these physical theories stop very abruptly at the Big Bang. We are left with very good experimental evidence for a Creation and a 'scientific' society with some considerable reluctance to discuss the possibility of a Creator. Consciousness research has no such limitation.

Consciousness appears to be evolving through these levels, with most of mankind now at the fourth level but advanced practitioners scouting out ahead and reporting back. The time taken to evolve through the various sub-levels of the first level (about 10^{10} years) seems enormous to us. But time itself, it turns out, is a construct of consciousness and varies with its level. We, all of us, have experience of this; 'normal' time and 'dream' time are quite different. This sort of thing I found to be one of the fascinating aspects of consciousness research – it is open to experimental checking by everyone. This is in considerable contrast to modern physics. If I wished to confirm the existence of the Ω^- particle I do not think I could do it. I could not get the money, or a big enough experimental team, or time on one of the few accelerators large enough. It is different in consciousness research.

Wilber likens the levels to the floors of a multi-storey building. He notes there are many things one can do on any one level: subdivide it into rooms in various ways, re-arrange the furniture, get new furniture, and so on. When one has exhausted not so much the possibilities of this (for these are endless) but our interest, one feels impelled to move up. The first kind of activity, confined to one floor, Wilber calls translation. The second is transformation, or transcendence. It seems to me that most research in physics has been mainly of the translatory sort, determining the laws of this lowest form of consciousness – matter. But some hints at the need for transcendence – the work on the fundamentals of quantum mechanics for instance. I think that the results of our quest for quarks are bringing this truth home to us.

At one time it seemed that the three quarks of the Zweig–Gell-Mann theory might be the 'fundamental buildings blocks' of the Universe. But, just like the chemical atoms and, in their turn the 'fundamental particles' of the 1940s, they proliferated. It now seems clear that they are *not* the fundamental building blocks of the Universe and furthermore, that no such things exist. The quantum-mechanical picture of one, whole and indivisible Universe is closer to reality. That picture makes the observer an essential part, suggests that observer and observed are one. To make further progress with this I think we need to transcend, to move up. And I think we are already on the move. The quest for quarks has led us to the need for transcendence.

Reading list

'Inside the Quark.' *Scientific American*, February 1981, pp. 64–8.

'Quarkonium.' E. D. Bloom and G. J. Feldman, *Scientific American*, May 1982, pp. 42–53.

The Tao of Physics. Fritjof Capra, Shambhala Press, Berkeley 1975.

The Forces of Nature. P. C. W. Davies, Cambridge University Press 1979.

'The Quantum Theory and Reality.' B. d'Espagnat, *Scientific American*, November 1979, pp. 128–140.

'A Unified Theory of Elementary Particles and Forces.' H. Georgi, *Scientific American*, April 1981, pp. 40–56.

'Quarks with Color and Flavor.' S. H. Glashow, *Scientific American*, October 1975, pp. 38–55.

Physics and Philosophy. W. Heisenberg, George Allen & Unwin, London 1959.

'Glueballs.' K. Ishikawa, *Scientific American*, November 1982, pp. 122–133.

'A Review of Quark Search Experiments.' L. W. Jones, *Review of Modern Physics*, October 1977, pp. 717–752.

The Interpretation of Nature and the Psyche. C. G. Jung and W. Pauli, Bollingen Series *LI*, Pantheon Books, New York 1959.

Man and his Symbols. C. G. Jung *et al.*, Picador 1978.

Atoms and Elements. D. M. Knight, Hutchinson, London 1967.

The Structure of Scientific Revolutions. T. S. Kuhn, University of Chicago Press 1970.

Quarks. C. B. A. McCusker, Encyclopaedia of Physics, 2nd Edition, Van Nostrand–Reinhold 1974, and 3rd Edition, Van Nostrand–Reinhold.

Mind and Matter. E. Schrödinger, Cambridge University Press 1959.

Up from Eden. Ken Wilber, Anchor-Doubleday, New York 1981.

Index

111719

539.721 M   111719

McCusker, Brian, 1919-

The quest for quarks